T0140522

Intelligent Systems Reference Library

Volume 61

Series editors

Janusz Kacprzyk, Polish Academy of Sciences, Warsaw, Poland
e-mail: kacprzyk@ibspan.waw.pl

Lakhmi C. Jain, University of Canberra, Canberra, Australia
e-mail: Lakhmi.Jain@unisa.edu.au

For further volumes:
http://www.springer.com/series/8578

About this Series

The aim of this series is to publish a Reference Library, including novel advances and developments in all aspects of Intelligent Systems in an easily accessible and well structured form. The series includes reference works, handbooks, compendia, textbooks, well-structured monographs, dictionaries, and encyclopedias. It contains well integrated knowledge and current information in the field of Intelligent Systems. The series covers the theory, applications, and design methods of Intelligent Systems. Virtually all disciplines such as engineering, computer science, avionics, business, e-commerce, environment, healthcare, physics and life science are included.

Animesh Adhikari · Jhimli Adhikari
Witold Pedrycz

Data Analysis and Pattern Recognition in Multiple Databases

 Springer

Animesh Adhikari
Parvatibai Chowgule College
Margao
India

Jhimli Adhikari
Narayan Zantye College
Bicholim
India

Witold Pedrycz
Department of Electrical and Computer
   Engineering
University of Alberta
Edmonton
Canada

ISSN 1868-4394
ISBN 978-3-319-37727-8
DOI 10.1007/978-3-319-03410-2
Springer Cham Heidelberg New York Dordrecht London

ISSN 1868-4408   (electronic)
ISBN 978-3-319-03410-2   (eBook)

Printed on acid-free paper

Springer is part of Springer Science+Business Media (www.springer.com)

To
Radha Ballav Adhikary, and
Manorama Adhikary

AA

Gouri Sankar Datta, and Rina Datta

JA

Ewa, Barbara, and Adam

WP

Foreword

I knew the authors of this book from reviewing their quality research papers and co-authored monograph many years ago. It is my honor to introduce you this new book concerning about multisourced data mining.

Analysis of big data from multiple sources is an emerging field of study. With the introduction of numerous channels of collecting data, analysis of multisourced data becomes a need of the hour. Several organizations such as banks, insurance companies, and shopping malls generate a huge amount of data coming from different sources. Quite commonly, data generated for a long period of time could be divided into a number of datasets to generate time/region specific knowledge.

A systematic analysis of multisourced data was started by S. Zhang, X. Wu, and C. Zhang in 2002. This field is gaining popularity as many people started working in this area by realizing its potential. A number of issues, such as big size of a database, presence of multiparty sensitive data, immovability of data from one place to another, and variety of data format originating from different sources have made data analysis and knowledge discovery more challenging.

Discovery of patterns and associations in multisourced data is a natural and important activity. Although classical patterns such as association rule, frequent itemset, and sequential pattern are of interest to study, patterns such as high frequency rule, heavy association rule, and exceptional pattern, specific to multidatabases, are also getting reported. They were selectively co-authored to the monograph "Developing Multi-Database Mining Applications" by A. Adhikari, P. Ramachandrarao, W. Pedrycz.

We understand that the classical data mining techniques have limitations in mining multiple large databases. Multisourced data mining techniques, such as local pattern analysis and pipelined feedback technique, provide a way to deal with multisourced large databases. We believe many specialized techniques will get reported to handle multisourced big databases.

Association rule mining (Chap. 2), global pattern discovery (Chap. 4), and mining patterns of select items (Chap. 6) provide different patterns discovery techniques in multisourced data. Some interesting item-based data analyzes are found in Chaps. 3 and 5. Interesting patterns, such as exceptional pattern, iceberg and periodic pattern are found in Chaps. 7–9, respectively. Chapter 10 discusses influence analysis between items. This book presents a recent research on mining multisourced data while highlighting contributions made in the past.

Over the time researchers have made efforts to discover techniques, approaches to deal with the new and challenging scenarios. Undoubtedly, in future, we will be witnessing new techniques to deal with massive, complex data originating from various sources.

China, April 2013   Shichao Zhang

Preface

Analysis of big data is a recent hot topic of research, and data analysis using multiple large databases is one such specific area. This area is gaining popularity, as the sources of multiple datasets are easily available due to significant technological advancements in capturing data and current requirements of our society.

We have noticed that multi-database mining is a different activity than mono-database mining. There are many challenges that one needs to tackle with. In many cases, data from multiple sources can not be moved to a single location. In this regard, the issues such as privacy-preservation, limited bandwidth offered by wireless channels, and retention of local features might play important roles. In these cases, local data are required to be mined locally.

Although mining multiple large databases poses many challenges, it offers many opportunities. Many decisions are based on multiple large databases located in different geographical regions. Such decisions would be more valid if they are based on the data distributed over the regions. First systematic studies[1] were carried out by Shichao Zhang, Xindong Wu and Chengqi Zhang. We continued investigations[2] on mining multiple large databases. The present book is based on the current research on knowledge discovery in multiple related databases.

The authors of this book extend their gratitude to Professor Lakhmi Jain for recognizing our work, and we thank Springer for accepting our work. We thank Ms. Varddhene for her excellent typesetting work. We also thank Ms. Ramya for overseeing the entire production process.

India   Animesh Adhikari
India   Jhimli Adhikari
Canada   Witold Pedrycz

---

[1] Zhang S, Zhang C, Wu X (2004) Knowledge discovery in multiple databases. Springer.

[2] Adhikari A, Ramachandrarao P, Pedrycz W (2010) Developing multi-databases mining applications. Springer.

Contents

Chapter 1
Introduction

Organizations that collect data from their multiple branches are common. Also, many established organizations possess data for a long period of time. Due to a spectrum of analyses, such data often need to be sub-divided into smaller databases. Thus, the number of multiple databases is increasing. Global decisions made by a multi-branch organization might be more appropriate if they were based on the data collected over the branches. In this chapter, we discuss various issues and approaches to multi-database mining. We look at distributed data mining. We discuss different data processing tasks that are often needed in a multi-database mining environment. We present various patterns and associations in multiple data sources and then briefly summarize chapters of this book. We also present here problems related to multi-database mining.

1.1 Motivation

In these days examples of multiple databases are not rare. In fact, the domain of multi-databases is expanding with the increasing needs of our civilization (Page and Craven 2003; Papadimitriou et al. 2005; Forestier et al. 2009; Adhikari and Rao 2013; Yan et al. 2006; Kargupta et al. 2000; Dzeroski 2003). Many governmental bodies such as meteorological and agricultural departments collect similar data coming from multiple sources. The central meteorological department collects data from different states while the state meteorological departments collect rainfall data from different parts of the state. Also, the central agricultural department collects data pertaining to crops from different regions of the country. Many organizations transact data for a long period of time, and accordingly the database grows over time. Such large databases might be divided into small databases for purpose of gaining time/region specific knowledge, and subsequently the time/region specific knowledge can be combined to derive global knowledge. Different branches of a national bank may store transactions centrally, and thus generate a large database. To gain further insights into knowledge acquired at the regional level, the bank might need to divide the database into regional (local) databases. Thus, the sources of multiple

A. Adhikari et al., *Data Analysis and Pattern Recognition in Multiple Databases*,
Intelligent Systems Reference Library 61, DOI: 10.1007/978-3-319-03410-2_1,
© Springer International Publishing Switzerland 2014

databases are common. Moreover, many large organizations transact from multiple branches. They also generate multiple databases. Thus, there is an urgent need to address problems originating from multiple data sources.

As the number of data collection channels increases and becomes more diversified, many real-world data mining tasks can easily acquire multiple databases from various sources. Can we perform data mining by amalgamating the databases originated from different sources? The answer to this question might not always be affirmative. If we have sensitive data originating from different sources we just cannot amalgamate them. Even if we are allowed to arrange the databases together, we might not always be able to mine it. It could be difficult to mine the combined database when it becomes too large. Multi-database mining techniques (Adhikari et al. 2010a; Wu and Zhang 2003; Adhikari and Rao 2008a) play important roles in these situations. A multi-database mining technique mines global patterns in multiple databases without amalgamating the local databases. While dealing with multiple databases, we might not be always interested in the global patterns. Sometimes local patterns could be used to find solutions to many problems. In this book we also present solutions to various problems using patterns in local databases.

Knowledge discovery in databases (KDD) is an important activity, and knowledge discovery in multiple databases (KDMD) falls under this realm. KDMD has received a lot of attention in the recent time (Zhang et al. 2004b; Adhikari et al. 2010b; Kargupta et al. 2008). There are numerous types of data in a distributed environment. In this book, we deal with multiple transactional databases. We discuss different patterns in multiple related databases. So, knowledge in multiple databases includes global patterns as well as knowledge synthesized using local patterns. For a multi-branch organization, an important decision could be based on the data distributed over the individual branches. A global decision might require an analysis of the entire data distributed over the branches. The validity of the decisions might depend on how effectively one can handle and comprehend relevant data at different branches. Other class of problems might require only to process local patterns, but not to synthesize global patterns.

Sometimes we need to divide a large database into smaller databases to carry out a data analysis, or to recognize relevant patterns. For example, a large temporal database could be divided into yearly databases, since a season re-appears on a yearly basis. Afterwards, one could make an analysis whether the yearly patterns are partial or fully periodic (Adhikari and Rao 2011). Moreover, many knowledge-driven data analyses could be performed on multiple databases (Adhikari et al. 2011a, b).

Discovering knowledge from multiple databases is an interesting yet highly challenging issue (Adhikari et al. 2010b). Some of the visible challenges are large size of a local database, variety of data format of the local databases, pattern synthesizing process and lack of systematic knowledge discovery techniques.

The chapter is organized as follows. In Sect. 1.2, we provide an overview of mining distributed databases. In Sect. 1.3, we discuss different approaches to multi-database mining techniques. Section 1.4 presents different data preprocessing techniques often required in a multi-database mining environment. Various patterns and associations are discussed in Sect. 1.5. Related works are covered in Sect. 1.6. We have mentioned the experimental settings adopted for different problems in Sect. 1.7. We discuss major domains related to multi-database mining and present some conclusions in Sect. 1.8.

1.2 Distributed Data Mining

Distributed data mining (DDM) deals with mining multiple databases distributed over different geographical regions. In the last few years, researchers have started addressing problems where the databases stored at different places cannot be moved to a central storage area for variety of reasons. In multi-database mining, there are no such restrictions. Thus, distributed data mining could be considered as a special type of multi-database mining. Distributed data mining environment often comes with different distributed sources of computation. The advent of ubiquitous computing (Greenfield 2006), sensor networks (Zhao and Guibas 2004), grid computing (Wilkinson 2009), and privacy-sensitive multiparty data (Kargupta et al. 2003) present examples where centralization of data is either not possible, or at least not always desirable.

There is no doubt that ubiquitous computing could form the next wave of computing. We experienced the first wave of computing due to the excessive use of mainframes in both academia and industries. Each mainframe is shared by lots of people. Now we are in the personal computing era, person and machine face at each other uncomfortably across the desktop. Moreover, a person sometimes needs to spend hours to finish the task. It makes a person tiresome. Next comes ubiquitous computing, or the age of *calm* technology, when technology recedes quietly into the background of our lives. As opposed to the desktop paradigm, in which a single user consciously engages a single device for a specialized purpose, someone using ubiquitous computing engages many computational devices and systems simultaneously, in the course of ordinary activities, and may not necessarily even be aware that they are doing so.

There are many domains where distributed processing of data becomes a natural and scalable solution. Distributed wireless applications define one of such important domain. Consider an ad hoc wireless sensor network where different sensor nodes are monitoring some time-critical events. Central collection of data from every sensor node may create heavy traffic over the limited bandwidth offered by wireless channels and this may also drain a lot of power from the individual devices. Apart from the issue of power consumption, DDM over wireless networks also requires an application to be run efficiently as many applications are time critical. The system might require monitoring and mining the on-board data stream generated by different sensors. Thus, centralization of databases is not desirable at all.

Many privacy-sensitive data mining adopt a distributed framework. The participating nodes exchange minimal amount of information without transmitting raw data. Stolfo et al. (1997) designed JAM system for mining multiparty distributed sensitive data such as financial fraud detection. Distributed data in health care, finance, counter-terrorism and homeland defense often use sensitive data held by different parties. This comes into direct conflict with an individual's need and right to privacy. Yi and Zhang (2007) have proposed a privacy-preserving distributed association rule mining protocol based on a semi-trusted mixer model. The protocol can protect the privacy of each distributed database against the coalition up to $n - 2$ other data sites or even the mixer if the mixer does not collude with any data site. Zhan et al. (2006) have proposed a secure protocol for multiple parties to collaboratively conduct association rule mining without disclosing their private data to each other or any other parties. Zhong (2007) has proposed algorithms for both vertically and horizontally partitioned data, with cryptographically strong privacy. The author presented two algorithms for vertically partitioned data; one of them reveals only the support count and the other reveals nothing. Inan et al. (2007) proposed methods for constructing the dissimilarity matrix of objects from different sites in a privacy preserving manner which can be used for privacy preserving clustering as well as database joins, record linkage and other operations that require pairwise comparison of individual private data objects horizontally distributed to multiple sites.

Industry, science, and commerce fields often need to analyze very large databases maintained over geographically distributed sites by using the computational power of distributed systems. Grid can play a significant role in providing an effective computational infrastructure support for this kind of data mining. Similarly, the advent of multi-agent systems has brought us a new paradigm for the development of complex distributed applications. During the past decades, there have been several models and systems proposed to apply agent technology to build distributed data mining. Through a combination of these two techniques, Luo et al. (2007) have investigated the different issues to build DDM on grid infrastructure and designed an agent grid intelligent platform as a test bed. Data mining algorithms and knowledge discovery processes are both compute and data intensive; therefore a grid can offer a computing and data management infrastructure for supporting decentralized and parallel data analysis. Congiusta et al. (2007) discussed how grid computing can be used to support distributed data mining. In this book, we deal with multiple transactional databases that are not necessarily sensitive. In the following section, we discuss how the existing approaches dealt with multiple large databases.

1.3 Multi-database Mining Approaches

We discuss three approaches to mining multiple large databases. In a distributed data mining environment, we may encounter different types of data. For example, stream data, geographical data, image data and transactional data are quite common. In this book, we deal with multiple transactional databases.

1.3.1 *Local Pattern Analysis*

Based on the number of data sources, patterns in multiple databases could be classified into three categories. They are local patterns, global patterns and patterns that are neither local nor global. A pattern based on a local database is called a local pattern. Local patterns are useful for local data analysis and decision making problems (Adhikari and Rao 2008b; Wu et al. 2005). On the other hand, global patterns are based on all the databases under consideration. They are useful for global data analyses (Adhikari and Rao 2008a; Wu and Zhang 2003). A convenient way to mine global patterns is to mine each local database, and then analyze all the local patterns to synthesize global patterns. This technique is simply called *local pattern analysis*. Zhang et al. (2003) designed local pattern analysis for the purpose of addressing various problems related to multiple large databases. Let there be n branches of a multi-branch company. Also, let D_i be the database corresponding to the i-th branch, $i = 1, 2, ..., n$. The essence of mining multiple databases using local pattern analysis is explained in Fig. 1.1 (Adhikari et al. 2010b).

Let LPB_i be the local pattern base corresponding to D_i, $i = 1, 2, ..., n$. In a multi-database environment, local patterns could be used in three ways by (1) analyzing local data, (2) synthesizing non-local patterns, and (3) expressing relevant statistics for decision making problems. Multi-database mining using local pattern analysis could be considered as an approximate method of mining multiple large databases. Thus, it might be required to enhance the quality of knowledge synthesized from multiple databases.

Adhikari et al. (2010a) viewed multi-database mining technique (MDMT) using local pattern analysis as a two-step process $\tau + \xi$, explained as follows:

- Mine each local database using a single database mining technique by applying a model τ (step 1)
- Synthesize patterns using an algorithm ξ (step 2)

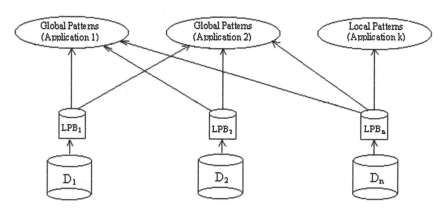

Fig. 1.1 Mining patterns in multiple databases with the use of local pattern analysis

We use notation MDMT: $\tau + \xi$ to represent a multi-database mining technique using local pattern analysis.

1.3.2 Sampling

In multi-database environment, the collection of all branch databases might be very large. Effective data analysis using a traditional data mining technique based on multi-gigabyte repositories has proven to be difficult. An approximate knowledge derived from large databases would be adequate for many decision support applications. Such applications might be necessary to help support decision-making in a rapid way. In these cases, one could tame multiple large databases by sampling (Babcock et al. 2003). For instance, a commonly used technique for approximate query answering is sampling (Cochran 1977). If an itemset is frequent in a large database then it is likely that the itemset is also frequent in a sample data. Thus, one could analyze approximately the database by analyzing the frequent itemsets in a representative sample data. A combination of sampling and local pattern analysis could be a useful technique for mining multiple databases for addressing many decision support applications.

1.3.3 Re-Mining

For the purpose of mining multiple databases, one could apply partition algorithm proposed by Savasere et al. (1995). The algorithm is designed for mining a very large database by partitioning. The algorithm works as follows. It scans the database twice. The database is divided into disjoint partitions, where each partition is small enough to fit in memory. In a first scan, the algorithm reads each partition and computes local frequency itemsets in each partition using apriori algorithm (Agrawal and Srikant 1994). In the second scan, the algorithm counts the supports of all local frequency itemsets toward the complete database. In this case, each local database could be considered as a partition. Though partition algorithm mines frequent itemsets in a database exactly, it might be an expensive solution to mining multiple large databases, since each database is required to be scanned twice. During the time of the second scanning, all the local patterns obtained at the first scan are analyzed. Thus, partition algorithm used for mining multiple databases could be considered as another type of local pattern analysis.

1.4 Pre-processing of Databases

Many application domains require collecting and processing of huge quantities of data coming from different sources. Examples include market basket data analysis, financial data analysis, scientific applications, and medical applications. However, it is difficult to identify universal procedures for analyzing data, since the characteristics of data might vary over the domains. In some cases, data obtained as series of values drawn on irregular time grids might be affected by noise. It requires the adoption of techniques customized on the basis of the context and on the goals of the analysis. In order to make the inspection and the processing of information easy and efficient, it is useful to transform the raw data into series of patterns that summarize their temporal evolution. This kind of analysis is usually preliminary or could be viewed as complementary to the execution of different tasks such as visual data exploration, data mining and summarization, knowledge extraction and reasoning. In view of achieving the final goal of the pattern and/or association extraction, we may need to perform some kind of data processing tasks. Data preprocessing is an important step for deriving meaningful knowledge. Such preprocessing may vary from application to application. In the following section we discuss a number of important tasks.

1.4.1 Preparation of Data Warehouses

Each branch database should be preprocessed to make data suitable for data mining. It could well be that all the data sources are not in the same format. Some times data need to be converted from one type to another. One needs to process it before any mining task takes place. A few important steps for preparing data at a branch are aggregation, sampling, dimensionality reduction, feature subset selection, feature creation, discretization and variable transformation (Tan et al. 2006). Relevant data are required to be retained for the purpose of mining. Also, the definitions of data are required to be the same at every data source. The preparation of data warehouse completed at every branch of the organization could be a significant task (Pyle 1999; Zhang et al. 2003). We have presented an extended model in Chap. 2 for synthesizing global patterns from local patterns in different data sources (Adhikari and Rao 2008a). A discussion is made on how this model could be used for mining different extreme association rules in multiple data sources. Also, it has been shown how the task of data preparation could be broken into sub-tasks so that the overall data preparation task becomes easier and can be realized in a systematic fashion. Although the model introduces many layers and interfaces for synthesizing global patterns, several of these layers and interfaces might not be required in a real-life application. Due to a heterogeneous nature of different data sources, data integration is often one of the most challenging tasks in managing modern information systems. Jiang et al. (2007) have proposed a framework for integrating multiple data sources when a single "best" value has to be chosen and stored for every attribute of an entity.

1.4.2 Temporal Aggregation

In historical databases, temporal aggregation is a process in which a time line is partitioned over time and the values of various attributes in the database are accumulated over these partitions (Dumas et al. 1998). A typical example of temporal aggregation is the monthly accumulation of salary payment. Due to the large variety of temporal data and their distribution over the time line, efficient algorithms to perform temporal grouping are required. Moon et al. (2003) proposed several methods for large-scale temporal aggregation. In this context choice of time granularity is an important issue as the characteristics of temporal patterns are heavily dependent on this parameter. Let us consider an online shop that acquires monthly reports from their web hosts. The web hosts deliver activity reports at regular intervals. Here time granularity refers to a month instead of a year. Therefore, for this application, month-wise time-stamped data could be accumulated to form smaller databases. Similarly, for stock market applications, weekly data accumulation may be preferred over monthly data. One of the reasons to summarize the time-stamped data is to reduce the amount of data. Afterwards, an effective algorithm could be designed to handle reduced data.

1.4.3 Partitioning Database

Partitioning a large database becomes an important operation in some cases. Consider an organization possessing data over fifty consecutive years. The organization might be interested in mining knowledge for various activities such as finding items whose supports are stable over the time (Adhikari et al. 2009), and extracting yearly periodic patterns (Adhikari and Rao 2011). In order to extract such knowledge, one could divide the given database into a number of yearly databases. Also for the purpose of mining change (Böttcher et al. 2008), one may require partitioning a given database. Thus, a single data source generates multiple databases. In a prediction problem one requires analyzing past events that would originate from the previous databases. Given such a problem, it might be strategically necessary to divide a large database into smaller databases. One could call these smaller databases as *time databases*. While dividing a large database, one requires selecting certain time period. Selection of time period is an important decision, and it is dependent on the problem. For many problems, one could divide the database into yearly databases (Adhikari et al. 2009, 2011b; Adhikari and Rao 2011). One needs to consider time granularity as one year for certain problems, since a season re-appears on a yearly basis and the customers' purchase patterns might vary from season to season.

1.4.4 Database Thinning

Database thinning refers to the process of discarding items in transactions that are not relevant to the given context. These items could be treated as outliers. Thinning process makes the size of transactions shorter and thus the mining process becomes easier. In order to estimate association between select items, or, to find patterns of select items in multiple databases, we may need to discard other items in a local database. One can apply database thinning to each local database. One could achieve this task by applying the certain steps (Adhikari et al. 2011a). First we apply an algorithm to partition a local database into two parts viz., forwarded database (FD) and remaining database. In the following paragraph, we discuss how to construct FD_i, from D_i, $i = 1, 2, ..., n$.

Initially, FD_i is kept empty. Let T_{ij} be the j-th transaction of D_i, $j = 1, 2, ..., |D_i|$. For D_i, a for-loop on j would run for $|D_i|$ times. At the j-th iteration, the transaction T_{ij} is tested. If T_{ij} contains at least one select item then FD_i is updated by $FD_i \cup \{T_{ij}\}$. At the end of the for-loop on j, FD_i gets constructed. After applying a thinning process, relevant patterns could be extracted. We study more about the application of thinning process in Chap. 6.

1.4.5 Ordering of Databases

In the context of mining multiple large databases, the order of mining each local database seems to be an important issue. As mentioned earlier, one can view MDMT as a two step process, where the second step i.e., synthesizing patterns might depend on the realization of the first step. We have presented an improved multi-database mining technique, called PFT, in Chap. 4. We see in Chap. 4 how the ordering of individual database mining plays an important role in enhancing the accuracy of global patterns.

1.4.6 Selection of Databases

An approximate form of knowledge resulting from large databases would be adequate for many decision support applications. In this sense, the selection of databases might be important in many decision support applications by reducing the cost of searching for necessary information. Their selection is based on the inherent knowledge residing in the database. For that purpose, one needs to mine each local database. Then we process the local patterns in different databases for the purpose of selecting relevant databases. Local patterns help selecting relevant databases. Based on local patterns, one can cluster the local databases. To answer the given query, one mines all the databases positioned in a relevant cluster.

In many cases, the clustering of databases is based on a measure of similarity between databases. Thus, a measure of similarity between two databases is an important design component whose development is based on local patterns present in the databases.

Wu et al. (2005) have proposed a similarity measure sim_1 to identify similar databases based on item similarity. The authors have designed an algorithm based on it to cluster databases for the purpose of selecting relevant databases. Such clustering is useful when the similarity is based on items present in different databases. This measure might not be useful for many multi-database mining applications where clustering of databases might be based on some other criteria. For example, if we are interested in the relevant databases based on transaction similarity then the above measures might not be appropriate. We proposed a measure $simi_1$ to find transaction similarity between two databases (Adhikari and Rao 2008b). We have designed a clustering algorithm based on $simi_1$ for the purpose of selecting relevant databases.

1.5 Patterns and Associations

When all the databases are allowed to be put together and it is possible to mine the combined database then there is no difference between multi-database mining and mono-database mining. The patterns returned from the combined database are known as global patterns. Global patterns are useful for global data analyses and global decision making problems. But, local patterns are extracted by mining a local database. Local patterns are sometimes useful for synthesizing global patterns as well as problems induced by local patterns.

Due to large size of a local database, multi-database mining becomes more challenging. Often one needs to apply an approximate method of mining multiple large databases by making use of local patterns. As a result we may encounter extreme types of patterns such as high frequency association rule (Wu and Zhang 2003), heavy association rules (Adhikari and Rao 2008a), and exceptional global patterns (Adhikari 2012) while mining multiple databases using local pattern analysis.

Since transaction identifiers (ids) are unique and would not usually be frequent, mining frequent patterns with transaction ids showing records they occurred in, provide an efficient way to mine frequent patterns in many types of databases including multiple tables and distributed databases. Ezeife and Zhang (2009) have proposed a set of algorithms TidFPs, for mining frequent patterns with their transaction ids in a single transaction database, in a multiple tables and distributed database.

Zhu and Wu (2007) have proposed DRAMA, a systematic framework for Discovering Relational patterns Across Multiple dAtabases. More specifically, given a series of data collections, authors tried to discover patterns coming from different databases with patterns' relationships satisfying the user specified constraints. The method sought to build a Hybrid Frequent Pattern tree (HFP-tree) from multiple databases, and mined patterns from the HFP-tree by integrating users' constraints into the pattern mining process.

Lan et al. (2007) have proposed a new kind of patterns, called rare utility itemsets, that consider not only individual profits and quantities, but also common existing periods and branches of items in a multi-database environment. Authors have also proposed a new mining approach, called Two-Phase Algorithm for Mining Rare Utility Itemsets in Multiple Databases (TP-RUI-MD), to discover rare utility itemsets.

Kum et al. (2006) have proposed an ApproxMAP algorithm, to mine approximate sequential patterns, called consensus patterns, from large sequence databases in two steps. First, sequences are organized into similarity groups, called clusters. Then, consensus patterns are mined directly from each cluster through multiple alignments.

A multi-domain sequential pattern is a sequence of events whose occurrence time is within a pre-defined time window. Given a set of sequence databases across multiple domains, Peng and Liao (2009) have aimed at mining multi-domain sequential patterns. Authors have proposed algorithm Naive in which multiple sequence databases are joined as one sequence database for utilizing traditional sequential pattern mining algorithms (e.g., PrefixSpan). Due to the nature of join operations, the algorithm Naive incurs substantial computing overhead. Later authors have proposed improved algorithms without any join operations for mining multi-domain sequential patterns.

A multi-branch company is often interested in high-frequency rules because they are supported by most of the branches for corporate profitability. Wu and Zhang (2003) have proposed a weighting model for synthesizing high frequency association rules from different data sources.

Zhong et al. (2001) proposed peculiarity rules as a new class of rules, which can be discovered from a relatively low number of peculiar data by searching the relevance among the peculiar data. Authors illustrated that such peculiarity rules represent a typically unexpected, interesting regularity hidden in databases.

Yan et al. (2006) have introduced a new paradigm, called ratio rule. Ratio rules are aimed at capturing the quantitative association knowledge. Authors have extended this framework to mining ratio rules from distributed and dynamic data sources. Authors have proposed an integrated method to mining ratio rules from distributed and changing data sources, by first mining the ratio rules from each data source separately through a novel, robust and adaptive one-pass algorithm, and then integrating the rules of each data source in a simple probabilistic model.

Zhang et al. (2009) have proposed a nonlinear method, named kernel estimation for mining global patterns (KEMGP), which adopts kernel estimation to synthesizing global patterns using local patterns. Authors also adopt a method to divide all the data in different databases according to attribute dimensionality, which reduces the total space complexity.

A global exceptional pattern describes interesting individuality of few branches. Therefore, it is interesting to identify such patterns. Adhikari (2012) and Zhang et al. (2004a) have introduced different strategies for identifying global exceptional patterns in multiple databases.

Principal component analysis (PCA) is frequently used for constructing the reduced representation of the data. The method often reduces the dimensionality of the original data by a large factor and constructs features that capture the maximally varying directions in the data. Kargupta et al. (2000) have proposed a technique of computing the collective principal component analysis from heterogeneous sites.

With regard to discovering patterns and associations in multiple databases, some contributions are presented in this book. An introduction to each chapter is given below.

An extended model of local pattern analysis is presented in Chap. 2. The notion of heavy association rule in multiple databases is introduced here. We present an algorithm for synthesizing heavy association rules in multiple databases. Also, we present the notion of exceptional association rule in multiple databases, and extend the algorithm to detect whether a heavy association rule is highly frequent or exceptional.

In Chap. 3, we introduce the notion of stability of an item. Based on the degree of stability of an item, we design an algorithm for clustering items in different data sources. We have presented the notion of best cluster by considering average degree of variation of a class. Also, an alternative algorithm is designed to find the best cluster among items in different data sources.

In Chap. 4, we present existing specialized as well as generalized techniques for mining multiple large databases. We formalize the idea of multi-database mining using local pattern analysis and present a new generalized technique for mining multiple large databases. It improves the quality of synthesized global patterns significantly.

Chapter 5 deals with the following issues. An algorithm for synthesizing the supports of the high frequency itemsets is presented here. Based on a measure of association, we synthesize association among items in a high frequency itemset. An algorithm for clustering local frequency items in multiple databases is presented.

One might be interested in a set of specific items in multiple databases. In Chap. 6, we present a model of mining global patterns of select items from multiple databases. A measure of overall association between two items in a database is presented. We have also extended the measure for a database whose transactions contain items along with the quantities purchased. An algorithm is presented based on the proposed measure for the purpose of grouping the frequent items in multiple databases.

In view of formalizing the idea of global exceptional patterns, we present a definition of global exceptional frequent itemset in multiple databases (Chap. 7). We present the notion of exceptional sources for a global exceptional frequent itemset, and present an algorithm for synthesizing global exceptional frequent itemsets.

Time stamped data are quite common. In Chap. 8, we present a new pattern, called notch, of an item in time-stamped databases. Based on a notch, we have presented two special kinds of notch, called generalized notch and iceberg notch, in time-stamped databases. An algorithm is presented for mining interesting icebergs in time-stamped databases.

We have extended the concept of certainty factor by incorporating support information for effective analysis of overlapped intervals. An improved algorithm is presented in Chap. 9 for identifying calendar-based periodic patterns. Based on the modified algorithm, we identify full as well as partial periodic calendar-based patterns.

Many data analyses are based on influence of items on other items. In Chap. 10, we present the notion of overall influence of a set of items on another set of items. Using this notion, we have presented two algorithms for analysing influence involving specific items in a database.

1.6 Related Studies

Enterprise applications usually involve huge, complex, and persistent data to work on, together with business rules and processes. In order to represent, integrate, and use information coming from distributed multiple sources, Hu and Zhong (2006) have presented a conceptual model with dynamic multi-level workflows corresponding to a mining-grid centric multi-layer grid architecture, for multi-aspect analysis in building an e-business portal on the Wisdom Web. The authors have showed that this integrated model would help to dynamically organize status-based business processes that govern enterprise application integration.

To reduce the cost of search in the data coming from all databases, we need to identify which databases are most likely relevant to a data mining application. For this purpose, Wu et al. (2005) have proposed an algorithm for selecting relevant databases. Adhikari and Rao (2008b) have proposed an algorithm for clustering a set of databases. Efficiency of the clustering process has been improved using the following strategies: reducing execution time of clustering algorithm, using more appropriate similarity measure, and storing frequent itemsets space efficiently. Yin and Han (2005) have proposed a new strategy for relational heterogeneous database classification.

Liu et al. (2001) have proposed multi-database mining technique that searches only the relevant databases. Identifying relevant databases is based on selecting the relevant tables (relations) that contain specific, reliable and statistically significant information pertaining to the query.

A general discussion on multi-database mining, applications, various issues and challenges can be found in Zhang et al. (2004b) and Adhikari et al. (2010b). Kargupta et al. (2004) have edited a book containing various issues on distributed data mining. Domingos (2003) surveyed some of the main drivers, problems and opportunities in multi-relational data mining. Wang et al. (2005) present various techniques in biological data mining and data management. The book also covers preprocessing tasks such as data cleaning and data integration as being applied to biological data.

1.7 Experimental Settings

We have carried out numerous experiments in different chapters to study the effectiveness of the proposed approaches. For Chaps. 2, 3, 4, 6, 8 and 10, all the experiments have been realized on a 1.6 GHz Pentium processor with 256 MB of memory using Visual C++ (version 6.0) software. For Chaps. 5 and 7, all the experiments have been implemented on a 2.8 GHz Pentium D dual core processor with 512 MB of memory using Visual C++ (version 6.0) compiler. Experiments in Chap. 9 have been implemented on a 2.4 GHz, core i3 processor with 4 GB of memory, running Windows 7 HB, using Visual C++ (version 6.0) software.

1.8 Conclusions

Mining and analysis of a single database has been thoroughly studied. Multi-database mining and analysis is applicable to many domains. A few examples are cited in Sect. 1.1. Mining and analysis of multiple data sources is a recent topic of data mining. Therefore, it needs further attention. In this book, we have confined our discussion on mining multiple large transactional databases. Two types of research articles mostly appeared just recently: (1) mining non-local patterns and (2) applications based on local as well as non-local patterns. In the following paragraphs we discuss a few areas that are potential sources of multi-databases mining applications.

As noted earlier, market basket data coming from multiple sources is an important domain. Market basket analysis is a useful method of discovering customer purchasing patterns by extracting associations or co-occurrences from stores' transactional databases. Because the information obtained from marketing, sales, services, and operation strategies, it has drawn increased research interest. The existing methods, however, may fail to discover important purchasing patterns in a multi-store environment, because of an implicit assumption that products under consideration are on shelf all the time across all stores. Chen et al. (2005) have proposed a new method to overcome this weakness and showed that the proposed method is computationally efficient, and that it has advantage over the traditional method when stores are diverse in size, product mix changes rapidly over time, and larger numbers of stores and periods are considered. Almost all patterns in a single data source can also be discovered in multiple data sources. Li et al. (2009) have proposed a method that produces some infrequent itemsets of potential interest by scanning multi-database frequent pattern tree, and extracts negative association rules of interest according to the proposed correlation model. Ling and Yang (2006) have proposed a novel method that predicts the classification of data from multiple sources without class labels in each source.

World Wide Web (WWW) is a large distributed repository of data. Su et al. (2006) have proposed a logical framework for identifying quality knowledge from different data sources. In many large e-commerce organizations, multiple data sources are often used to describe the same customers, thus it is important to consolidate data of multiple sources for intelligent business decision making.

The popularity of the Internet as well as the availability of powerful computers and high-speed network technologies as low-cost commodity components is changing the way we use computers today. These technology opportunities have led to the possibility of using distributed computers as a single, unified computing resource, leading to what is popularly known as Grid computing (Foster and Kesselman 1999). Clusters and grids of workstations provide available resources for data mining processes. To exploit these resources, new distributed algorithms are necessary, particularly concerning the way to distribute data and to use this partition. Fiolet and Toursel (2007) have presented a clustering algorithm known as distributed progressive clustering, for providing an "intelligent" distribution of data on grids. Cluster and grid computing will be playing a dominant role in the next generation of computing. In a distributed environment, a large database could be fragmented vertically and/or horizontally. This might bring additional complexities for mining patterns in multiple large databases. Agrawal and Shafer (1999) introduced a parallel version of apriori algorithm.

Distributed data mining for wireless applications is another active area of multi-database mining. Challenges here are somewhat different from that of classical multi-database mining. Bandwidth limitation is one of the major constraints in this domain. There are other constraints, such as power consumption. The next generation algorithms will have to deal with these important constraints.

Data privacy is likely to remain an important issue in data mining research and applications. The field of privacy-preserving data mining has started recently. Da Silva and Klusch (2006) have proposed KDEC-S algorithm for distributed data clustering, which is shown to provide mining results while preserving confidentiality of original data. Stankovski et al. (2008) have designed *DataMiningGrid* system to meet the requirements of modern and distributed data mining scenarios. Based on the Globus Toolkit and other open technology and standards, the *DataMiningGrid* system provides tools and services facilitating the grid-enabling of data mining applications without any intervention on the application side. In future, the concepts and various issues will get formalized. Ashok and Mukkamala (2011) have proposed a new distributed data mining approach where each data owner derives association rules locally, sanitizes them if necessary, and sends them to a third-party data miner. The data miner collects local rules from all data owners, regenerates an estimate of global data, and performs global data mining. More privacy-preserving algorithms are likely to appear as more applications on privacy-sensitive data are likely to emerge in the future.

Multi-agent systems (MAS) offer architecture for distributed problem solving. DDM algorithms focus on one class of such distributed problem solving tasks, analysis and modeling of distributed data. Da Silva et al. (2005) offer a perspective on DDM algorithms in the context of multi-agents systems. It discusses broadly the connection between DDM and MAS. In future, many DDM algorithms are likely to come in association with MAS.

With the increasing popularity of object-oriented database systems in advanced database applications, it is also important to study the data mining methods in object-oriented data. Han et al. (1998) investigated issues on generalization-based data mining in object-oriented databases considering three crucial aspects: (1) generalization of complex objects, (2) class-based generalization and (3) extraction of different kinds of rules. The authors proposed an object cube model for class-based generalization, on-line analytical processing and data mining. Various issues of multiple object-oriented databases deserve to be investigated.

Biological databases contain a wide variety of data types, often with rich relational structure. Consequently multi-relational data mining techniques frequently are applied to biological data. Page and Craven (2003) have presented several applications of multi-relational data mining to biological data, taking care to cover a broad range of multi-relational data mining techniques. The field of bioinformatics is expanding rapidly. In this field large multiple as well as complex relational tables are dealt with frequently.

Clinical laboratory databases are among the largest generally accessible, detailed records of human phenotype. They will likely have an important role in future studies designed to tease out associations between human gene expression, presentation and progression of disease. Multi-database mining will be playing an important role in this area (Siadaty and Harrison 2008).

The significant increase in the availability of massive, complex data from various sources is creating computing, storage, communication, and human-computer interaction challenges for data mining. Providing a framework to better understand these fundamental issues, Kargupta et al. (2008) have surveyed promising approaches to data mining problems that span an array of disciplines. In the coming years, we will be witnessing more applications of multi-databases mining as well as integration of database system (Ozsu and Valduriez 2011) and multi-database mining technologies.

References

Adhikari A (2012) Synthesizing global exceptional patterns in different data sources. J Intell Syst 21(3):293–323

Adhikari A, Rao PR (2008a) Synthesizing heavy association rules from different real data sources. Pattern Recogn Lett 29(1):59–71

Adhikari A, Rao PR (2008b) Efficient clustering of databases induced by local patterns. Decis Support Syst 44(4):925–943

Adhikari J, Rao PR (2013) Identifying calendar-based periodic patterns. In: Ramanna S, Jain L, Howlett RJ (eds) Emerging paradigms in machine learning, pp 329–357. Springer, Berlin

Adhikari J, Rao PR, Adhikari A (2009) Clustering items in different data sources induced by stability. Int Arab J Inf Technol 6(4):394–402

Adhikari A, Ramachandrarao P, Prasad B, Adhikari J (2010a) Mining multiple large data sources. Int Arab J Inf Technol 7(2):241–249

Adhikari A, Ramachandrarao P, Pedrycz W (2010b) Developing multi-databases mining applications. Springer, London

Adhikari A, Ramachandrarao P, Pedrycz W (2011a) Study of select items in different data sources by grouping. Knowl Inf Syst 27(1):23–43

Adhikari J, Rao PR, Pedrycz W (2011b) Mining icebergs in time-stamped databases. In: Proceedings of Indian international conferences on artificial intelligence, pp 639–658

Agrawal R, Shafer J (1999) Parallel mining of association rules. IEEE Trans Knowl Data Eng 8(6):962–969

Agrawal R, Srikant R (1994) Fast algorithms for mining association rules. In: Proceedings of international conference on very large data bases, pp 487–499

Ashok VG, Mukkamala R (2011) Data mining without data: a novel approach to privacy-preserving collaborative distributed data mining. Workshop on Privacy in the Electronic Society, pp 159–164

Babcock B, Chaudhury S, Das G (2003) Dynamic sample selection for approximate query processing. In: Proceedings of ACM SIGMOD conference management of data, pp 539–550

Böttcher M, Hoppner F, Spiliopoulou M (2008) On exploiting the power of time in data mining. SIGKDD Explor 10(2):3–11

Chen Y-L, Tang K, Shen R-J, Hu Y-H (2005) Market basket analysis in a multiple store environment. Decis Support Syst 40(2):39–354

Cochran WG (1977) Sampling techniques, 3rd edn. Wiley, New York

Congiusta A, Talia D, Trunfio P (2007) Service-oriented middleware for distributed data mining on the grid. J Parallel Distrib Comput 68(1):3–15

Da Silva JC, Klusch M (2006) Inference in distributed data clustering. Eng Appl Artif Intell 19(4):363–369

Da Silva JC, Giannellab C, Bhargava R, Kargupta H, Klusch M (2005) Distributed data mining and agents. Eng Appl Artif Intell 18(7):791–807

Domingos P (2003) Prospects and challenges for multi-relational data mining. SIGKDD Explor 5(1):80–83

Dumas M, Fauvet MC, Scholl PC (1998) Handling temporal grouping and pattern-matching queries in a temporal object model. In: Proceedings of CIKM, pp 424–431

Dzeroski S (2003) Multi-relational data mining: an introduction. SIGKDD Explor 5(1):1–16

Ezeife CI, Zhang D (2009) TidFP: mining frequent patterns in different databases with transaction ID. In: Proceedings of DaWaK, pp 125–137

Fiolet V, Toursel B (2007) A clustering method to distribute a database on a grid. Future Gener Comput Syst 23(8):997–1002

Forestier G, Wemmert C, Pierre Gançarski P, Inglada J (2009) Mining multiple satellite sensor data using collaborative clustering. In: Proceedings of ICDM workshops, pp 501–506

Foster I, Kesselman C (eds) (1999) The grid: blueprint for a future computing infrastructure. Morgan Kaufmann, San Francisco

Greenfield A (2006) Everyware: the dawning age of ubiquitous computing, 1st edn. New Riders Publishing, Indianapolis

Han J, Nishio S, Kawano H, Wang W (1998) Generalization-based data mining in object-oriented databases using an object cube model. Data Knowl Eng 25(1–2):55–97

Hu J, Zhong N (2006) Organizing multiple data sources for developing intelligent e-business portals. Data Min Knowl Disc 12(2–3):127–150

Inan A, Kaya SV, Saygın Y, Savas E, Hintoglu AA, Levi A (2007) Privacy preserving clustering on horizontally partitioned data. Data Knowl Eng 63(3):646–666

Jiang Z, Sarkar S, De P, Dey B (2007) A framework for reconciling attribute values from multiple data sources. Manage Sci 53(12):1946–1963

Kargupta H, Huang W, Krishnamurthy S, Park B, Wang S (2000) Collective PCA from distributed and heterogeneous data. In: Proceedings of the 4th European conference on principles and practice of knowledge discovery in databases, pp 452–457

Kargupta H, Liu K, Ryan J (2003) Privacy sensitive distributed data mining from multi-party data. In: Proceedings of intelligence and security informatics, pp 336–342

Kargupta H, Joshi A, Sivakumar K, Yesha Y (2004) Data mining: next generation challenges and future directions. MIT Press, Cambridge

Kargupta H, Han J, Yu PS, Motwani R, Kumar V (2008) Next generation of data mining. Springer, Berlin

Kum H-C, Chang HC, Wang W (2006) Sequential pattern mining in multi-databases via multiple alignment. Data Min Knowl Disc 12(2–3):151–180

Lan G-C, Hong T-P, Tseng VS (2007) A novel algorithm for mining rare-utility itemsets in a multi-database environment. In: Proceedings of the 26th workshop on combinatorial mathematics and computation theory, pp 293–302

Li H, Shen Y, Hu X (2009) A novel mining method of global negative association rules in multi-database. In: Proceedings of IEEE international conference on intelligent computing and intelligent systems, pp 392–396

Ling CX, Yang Q (2006) Discovering classification from data of multiple sources. Data Min Knowl Disc 12(2–3):181–201

Liu H, Lu H, Yao J (2001) Toward multi-database mining: identifying relevant databases. IEEE Trans Knowl Data Eng 13(4):541–553

Luo J, Wang M, Hu J, Shi J (2007) Distributed data mining on agent grid: issues, platform and development toolkit. Future Gener Comput Syst 23(1):61–68

Moon B, Lopez IFV, Immanuel V (2003) Efficient algorithms for large-scale temporal aggregation. IEEE Trans Knowl Data Eng 15(3):744–759

Ozsu MT, Valduriez P (2011) Principles of distributed database systems. Springer, Berlin

Page D, Craven M (2003) Biological applications of multi-relational data mining. SIGKDD Explor 5(1):69–79

Papadimitriou S, Sun J, Faloutsos C (2005) Streaming pattern discovery in multiple time-series. In: Proceedings of VLDB, pp 697–708

Peng W-C, Liao Z-X (2009) Mining sequential patterns across multiple sequence databases. Data Knowl Eng 68(10):1014–1033

Pyle D (1999) Data preparation for data mining. Morgan Kaufmann, San Francisco

Savasere A, Omiecinski E, Navathe S (1995) An efficient algorithm for mining association rules in large databases. In: Proceedings of the 21st international conference on very large data bases, pp 432–443

Siadaty MS, Harrison JH Jr (2008) Multi-database mining. Clin Lab Med 28(1):73–82

Stankovski V, Swain M, Kravtsov V, Niessen T, Wegener D, Kindermann J, Dubitzky W (2008) Grid-enabling data mining applications with DataMiningGrid: an architectural perspective. Future Generation Computer Systems 24(4):259–279

Stolfo S, Prodromidis AL, Chan PK (1997) JAM: java agents for meta-learning over distributed databases. In: Proceedings of 3rd international conference on knowledge discovery and data mining, pp 74–81

Su K, Huang H, Wu X, Zhang S (2006) A logical framework for identifying quality knowledge from different data sources. Decis Support Syst 42(3):1673–1683

Tan P-N, Kumar V, Steinbach M (2006) Introduction to data mining. Pearson Education, Boston

Wang JT, Zaki MJ, Toivonen HT, Shasha DE (2005) Data mining in bioinformatics. Springer, London

Wilkinson (2009) Grid computing: techniques and applications. CRC Press, Boca Raton

Wu X, Zhang S (2003) Synthesizing high-frequency rules from different data sources. IEEE Trans Knowl Data Eng 14(2):353–367

Wu X, Zhang C, Zhang S (2005) Database classification for multi-database mining. Inf Syst 30(1):71–88

Yan J, Liu N, Yang Q, Zhang B, Cheng Q, Chen Z (2006) Mining adaptive ratio rules from distributed data sources. Data Min Knowl Disc 12(2–3):249–273

Yi X, Zhang Y (2007) Privacy-preserving distributed association rule mining via semi-trusted mixer. Data Knowl Eng 63(2):550–567

Yin X, Han J (2005) Efficient classification from multiple heterogeneous databases. In: Proceedings of 9th European conference on principles and practice of knowledge discovery in databases, pp 404–416

Zhan J, Matwina S, Chang LW (2006) Privacy-preserving collaborative association rule mining. J Netw Comput Appl 30(3):1216–1227

Zhang S, Wu X, Zhang C (2003) Multi-database mining. IEEE Comput Intell Bull 2(1):5–13

Zhang C, Liu M, Nie W, Zhang S (2004a) Identifying global exceptional patterns in multi-database mining. IEEE Comput Intell Bull 3(1):19–24

Zhang S, Zhang C, Wu X (2004b) Knowledge discovery in multiple databases. Springer, London

Zhang S, You X, Jin Z, Wu X (2009) Mining globally interesting patterns from multiple databases using kernel estimation. Expert Syst Appl 36(8):10863–10869

Zhao F, Guibas L (2004) Wireless sensor networks: an information processing approach. Morgan Kaufmann, San Francisco

Zhong S (2007) Privacy-preserving algorithms for distributed mining of frequent itemsets. Inf Sci 177(2):490–503

Zhong N, Yao YY, Ohshima M, Ohsuga S (2001) Interestingness, peculiarity, and multi-database mining. In: Proceedings of ICDM, pp 566–576

Zhu X, Wu X (2007) Discovering relational patterns across multiple databases. In: Proceedings of ICDE, pp 726–735

Chapter 2
Synthesizing Different Extreme Association Rules from Multiple Databases

The model of local pattern analysis provides sound solutions to many multi-database mining problems. In this chapter we discuss different types of extreme association rules in multiple databases viz., heavy association rule, high-frequency association rule, low-frequency association rule, and exceptional association rule. Also, we show, how one can apply the model of local pattern analysis systematically and effectively. For this purpose, we present an extended model of local pattern analysis. The extended model has been applied to mine heavy association rules in multiple databases. Also, we justify why the extended model works more effectively. An algorithm for synthesizing heavy association rule in multiple databases is presented. Furthermore, we show that the algorithm identifies whether a heavy association rule is high-frequency rule or exceptional rule. Experimental results are provided for both synthetic and real-world datasets and a detailed error analysis is carried out. Furthermore, we present a comparative analysis by contrasting the proposed algorithm with some of those reported in the literature. This analysis is completed by taking into consideration the criteria of execution time and average error.

2.1 Introduction

In the Chap. 1, the limitations of using "conventional" data mining technique for mining multiple large databases have been discussed. In many decision support applications, an approximate knowledge stemming from multiple large databases might result in significant savings when being used in decision-making. Hence the model of local pattern analysis (Zhang et al. 2003) used for mining multiple large databases can constitute a viable solution. In this chapter, we show how one could apply the model of local pattern analysis in a systematic and efficient manner for mining different types of extreme association rules in multiple databases.

The analysis of relationships existing among variables is a fundamental task positioned at the heart of many data mining problems. Mining association rules has received a lot of attention to the data mining community. For instance, an

A. Adhikari et al., *Data Analysis and Pattern Recognition in Multiple Databases*, Intelligent Systems Reference Library 61, DOI: 10.1007/978-3-319-03410-2_2, © Springer International Publishing Switzerland 2014

association rule expresses how the purchase of a group of items, called an *itemset*, affects the purchase of another group of items. Association rule mining is based on two measures quantifying the quality of the rules, that is support (*supp*) and confidence (*conf*); see Agrawal et al. (1993). An association rule r in database DB can be expressed symbolically as $X \rightarrow Y$, where X and Y are two itemsets in database DB. It expresses an association between the itemsets X and Y, called the antecedent and consequent of r, respectively. The meaning attached to this type of implication could be clarified as follows. If the items in X are purchased by a customer then the items in Y are likely to be purchased by the same customer at the same time. The interestingness of an association rule could be expressed by its support and confidence. Let E be a Boolean expression defined on the items in DB. Support of E in DB is defined as the fraction of transactions in DB such that the Boolean expression E is true for each of these transactions. We denote the support of E in DB as $supp_a(E, DB)$. The support and confidence of association rule r is expressed as follows:

$$supp_a(r, DB) = supp_a(X \cap Y, DB), \text{and}$$
$$conf_a(r, DB) = supp_a(X \cap Y, DB)/supp_a(X, DB)$$

Later, we shall be dealing with synthesized support and synthesized confidence of an association rule. Thus, it is required to differentiate between actual support/ confidence with synthesized support/confidence of an association rule. The subscript a used in the notation of support/confidence for referring the actual support/ confidence of an association rule. On the other hand, the subscript s in the notation of support/confidence is used to refer synthesized support/confidence of an association rule. A synthesized support/confidence of an association rule might depend on the technique applied to synthesizing support/confidence. We present here a technique for synthesizing support and confidence of an association rule in multiple databases. We say that an association rule r in database DB is *interesting* if the following relationships hold

$$supp_a(r, DB) \geq minimum\ support(\alpha), \text{and}$$
$$conf_a(r, DB) \geq minimum\ confidence(\beta)$$

The values of the parameters α and β are user-defined. The collection of association rules extracted from a database for the given values of α and β is called a *rulebase*.

In a multi-database mining environment, often one needs to handle multiple large databases. As a result, one may come across various types of patterns. Association rule mining (Agrawal et al. 1993) is an important and popular data mining task. It has many applications to different areas of computing (Zhang and Wu 2011). In this chapter, we are interested in mining association rules in multiple databases that are extreme in some sense. These association rules are induced by different data sources, and thus, these rules are specific to multi-database mining environment. An association rule in multiple databases becomes more interesting if it possesses higher support and higher confidence. This type of association rules is called heavy association rules (Adhikari and Rao 2008). Sometimes the number of times an association rule gets reported from local databases becomes an interesting issue. In the context of multiple databases, an association rule is called high-frequency rule (Wu and Zhang 2003) if it is extracted from many databases. In this context an association rule is called low-frequency rule (Adhikari and Rao 2008) if it is extracted from a few databases. Some association rules possess high support but have been extracted from a few databases only. These association rules are called exceptional association rules (Adhikari and Rao 2008). Many corporate decisions could be influenced by these types of extreme association rules in multiple databases. Thus, it is important to mine them. In the next section, we present different extreme association rules, and then we present a model of mining such association rules.

The chapter is organized as follows. We discuss some "extreme" types of association rules (Sect. 2.2). In Sect. 2.3, we present the problem formally. An extended model of local pattern analysis is presented in Sect. 2.4. We discuss related work in Sect. 2.5. We present an algorithm for synthesizing different extreme association rules in Sect. 2.6. In this section, we have also defined error of the experiment. We present experimental result in Sect. 2.7. Finally, some conclusions are provided in Sect. 2.8.

2.2 Some Extreme Types of Association Rule in Multiple Databases

Consider a large company with transactions originating from n branches. Let D_i be the database corresponding to the i-th branch of this multi-branch company, $i = 1, 2, \ldots, n$. Furthermore let D be the union of all branch databases. First, we define a heavy association rule in a single database. Afterwards, we define a heavy association rule in multiple databases.

Definition 2.1 An association rule r in database DB is heavy if $supp_a(r, DB) \geq \mu$, and $conf_a(r, DB) \geq v$, where μ $(>\alpha)$ and v $(>\beta)$ are the user-defined thresholds of high-support and high-confidence for identifying heavy association rules in DB, respectively. •

If an association rule is heavy in a local database then it might not be heavy in D. An association rule in D might have different statuses in different local databases. For example, it might be a heavy association rule, or an association rule, or a suggested association rule (defined later), or absent in a local database. Thus, we need to synthesize an association rule for determining its overall status in D. The method of synthesizing an association rule is discussed in Sect. 2.6. After synthesizing an association rule, we get its synthesized support and synthesized confidence in D. Let $supp_s(r, DB)$ and $conf_s(r, DB)$ denote synthesized support and synthesized confidence of association rule r in DB, respectively. A heavy association rule in multiple databases is defined as follows:

Definition 2.2 Let D be the union of all local databases. An association rule r in D is heavy if $supp_s(r, D) \geq \mu$, and $conf_s(r, D) \geq v$, where μ and v are the user-defined thresholds of high-support and high-confidence used for identifying heavy association rules in D, respectively. •

Apart from synthesized support and synthesized confidence of an association rule, the frequency of an association rule is an important issue in multi-database mining. We define *frequency* of an association rule as the number of extractions of the association rule from different databases. If an association rule is extracted from k out of n databases then the frequency of the association rule is k, $0 \leq k \leq n$. An association rule may be high-frequency rule or, low-frequency rule, or neither high-frequency rule nor low-frequency rule in multiple databases. We could arrive in such a conclusion only if we have user-defined thresholds of low-frequency (γ_1) and high-frequency (γ_2) of an association rule, for $0 < \gamma_1 < \gamma_2 \leq 1$. A low-frequency association rule is extracted from less than $n \times \gamma_1$ databases. On the other hand, a high-frequency association rule is extracted from at least $n \times \gamma_2$ databases. In the context of multi-database mining using local pattern analysis, we define a high-frequency association rule and a low-frequency association rule as follows:

Definition 2.3 Let an association rule be extracted from k out of n databases. Then the association rule is low-frequency rule if $k < n \times \gamma_1$, where γ_1 is the user-defined threshold of low-frequency. •

Definition 2.4 Let an association rule be extracted from k out of n databases. Then the association rule is high-frequency rule if $k \geq n \times \gamma_2$, where γ_2 is the user-defined threshold of high-frequency. •

While synthesizing heavy association rules in multiple databases, it may be worth noting some other attributes of a synthesized association rule. For example, high-frequency, low-frequency, and exceptionality are interesting as well as important attributes of a synthesized association rule. We have already defined high-frequency association rule and low-frequency association rule in multiple databases. We now define an exceptional association rule in multiple databases:

Definition 2.5 A heavy association rule in multiple databases is exceptional if it is a low-frequency rule. •

It may be worth contrasting a heavy association rule, a high-frequency association rule with an exceptional association rule in multiple databases.

- An exceptional association rule is also a heavy association rule.
- A high-frequency association rule is not an exceptional association rule, and vice versa.
- A high-frequency association rule is not necessarily be a heavy association rule.
- There may exist heavy association rules that are neither high-frequency rule nor exceptional rule.

The goal of this chapter is to extract these extreme association rules from multiple databases. For this purpose, we present an extended model of local pattern analysis.

2.3 Problem Statement

In the previous section, we learnt different types of extreme association rules. As discussed in Chap. 1, we have observed some difficulties in extracting different extreme association rules in the union of all branch databases by employing a traditional data mining technique. Therefore, we synthesize different extreme association rules by using patterns in branch databases. Let D be the union of all branch databases. Also, let RB_i and SB_i be the rulebase and suggested rulebase corresponding to database D_i, respectively. An association rule $r \in RB_i$, if $supp_a$ $(r, D_i) \geq \alpha$, and $conf_a(r, D_i) \geq \beta$, $i = 1, 2,..., n$. An association rule $r \in SB_i$, if $supp_a(r, D_i) \geq \alpha$, and $conf_a(r, D_i) < \beta$. There is a tendency of a suggested association rule in a database to become an association rule in another database. Apart from the association rules, we also consider the suggested association rules for synthesizing heavy association rules in D. The reasons for considering suggested association rules are given as follows. Firstly, we could synthesize support and confidence of an association rule in D more accurately. Secondly, we could synthesize high-frequency association rules in D more accurately. Thirdly, some experimental results have shown that the number of suggested association rules could be significant for some databases. In general, the accuracy of synthesizing an association rule increases as the number of extractions of the association rule increases. Thus, we consider suggested association rules also in synthesizing heavy association rules in D. In addition, the number of transactions in a database would be required in synthesizing an association rule. We define *size* of database DB as the number of transactions in DB, denoted by $size(DB)$. We state the problem as follows.

Let there be n databases $D_1, D_2,..., D_n$. Let RB_i and SB_i be the set of association rules and suggested association rules in D_i, respectively, $i = 1, 2,..., n$. Synthesize heavy association rules in the union of all databases (D) based on RB_i and SB_i, $i = 1, 2,..., n$. Also, notify whether each heavy association rule is high-frequency rule or exceptional rule in D.

2.4 An Extended Model of Local Pattern Analysis for Synthesizing Global Patterns

Let D_i be the database corresponding to i-th branch of the organization, $i = 1, 2,..., n$. Patterns in multiple databases could be grouped into the following categories based on the number of databases: local patterns, global patterns, and patterns that are neither local nor global. A pattern based on a branch database is called a *local pattern*. On the other hand, a *global pattern* is based on all databases under consideration. An essence of the extended model of local pattern analysis (Adhikari and Rao 2008) is illustrated in Fig. 2.1. The extended model comes with a set of interfaces and a set of layers. Each interface realizes a set of operations and produces dataset(s) (or, knowledge) based on the dataset(s) available at the next lower layer. There are four interfaces of the proposed model of synthesizing global patterns from local patterns.

Interface 2/1 is concerned with different operations on data realized at the lowest layer. By applying these operations, we come up with a processed database resulting from a local (original) database. These operations are performed on each branch database. Interface 3/2 applies a filtering algorithm to each processed database to separate relevant data from possible outliers. In particular, if we are interested in studying durable items then the transactions containing only non-durable items could be treated as outlier transactions. Different interesting criteria could be set to filter data. This interface supports loading data into the respective data warehouse. Interface 4/3 mines (local) patterns in each local data warehouse. There are two types of local patterns: local patterns and suggested local patterns. A suggested local pattern is close but fails to fully satisfy the requisite interestingness criteria. The reasons for considering suggested patterns are given as follows. Firstly, by admitting these patterns, we could synthesize patterns more accurately. Secondly, due to the stochastic nature of the transactions, the number of suggested patterns could be significant in some databases. Thirdly, there is a tendency that a suggested pattern of one database could become a local pattern in some other

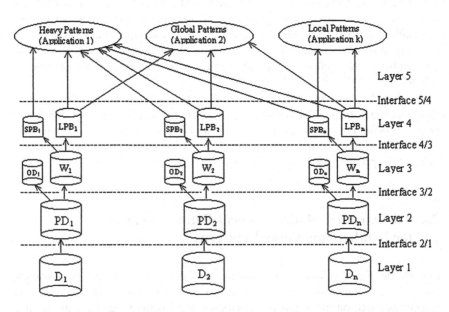

Fig. 2.1 A model of synthesizing global patterns from local patterns in different databases

databases. Thus, the correctness of synthesizing global patterns would increase as the number of local patterns increases. Therefore, the extended model becomes effective in synthesizing non-local patterns. Consider a multi-branch company having n databases. Let LPB_i and SPB_i be the local pattern base and suggested local pattern base corresponding to i-th branch of the organization, respectively, $i = 1, 2,..., n$. Interface 5/4 synthesizes global patterns, or analyses local patterns in order to find solutions to many problems.

At the lowest layer, all the local databases are retained. We may need to process these databases for the purpose of data mining task. Various data preparation techniques (Pyle 1999)—data preprocessing like data cleaning, data transformation, data integration, and data reduction are applied to data in the local databases. We get the processed database PD_i corresponding to the original database D_i, $i = 1, 2,..., n$. Then we retain all the data that are relevant to the data mining applications. Using a relevance analysis, one could detect outlier data (Last and Kandel 2001) from processed database. A relevance analysis is dependent on the context and varies from one application to another. Let OD_i be the outlier database corresponding to the i-th branch, $i = 1, 2,..., n$. Sometimes these databases are also used in some other applications. After removing outliers from the processed database we form data warehouse, and the data present there become ready for data mining task. Let W_i be the data warehouse corresponding to i-th branch. Local patterns for the i-th branch are extracted from W_i, $i = 1, 2,..., n$. Finally, the local patterns are forwarded to the central office for synthesizing global patterns, or completing analysis of local patterns. Many data mining applications could be developed based on the local patterns in different databases. In particular, if we are interested in synthesizing global frequent itemsets then an itemset may not be extracted from all the databases. It might be required to estimate the support of a frequent itemset in a database that fails to report it. Thus, in essence, a global frequent itemset synthesized from local frequent itemsets is approximate. If any one of the local databases is too large to apply a traditional data mining technique then this model would fail. In this situation, one could apply an appropriate sampling technique to reduce the size of the corresponding local database. Otherwise, the database could be partitioned into sub-databases. As a result, the error associated with the results produced in data analysis would increase.

Though the above model introduces many layers and interfaces for synthesizing global patterns, in a real life application, some of these layers might not be fully exploited. In this chapter, we discuss a problem of multi-database mining that uses the above model.

2.5 Related Work

Some applications of multiple large databases have been discussed in Chap. 1. Association rule mining gives rise to interesting association between two itemsets in a database. The notion of association rule was introduced by Agrawal et al.

(1993). The authors have proposed an algorithm to mine frequent itemsets in a database. Many algorithms to extract association rules have been reported in the literature. In what follows, we present a few interesting algorithms for extracting association rules in a database. Agrawal and Srikant (1994) have proposed apriori algorithm that uses breadth-first search strategy to count the supports of itemsets. The algorithm uses an improved candidate generation function, which exploits the downward closure property of support and makes it more efficient than earlier algorithm. Han et al. (2000) have proposed data mining method of FP-growth (frequent pattern growth) which uses an extended prefix-tree (FP-tree) structure to store the database in a compressed form. FP-growth adopts a divide-and-conquer approach to decompose both the mining tasks and databases. It uses a pattern fragment growth method to avoid the costly process of candidate generation and testing. Savasere et al. (1995) have introduced partition algorithm. The database is scanned only twice. In the first scan, the database is partitioned and in each partition support is counted. Then the counts are merged to generate potential frequent itemsets. In the second scan, the potential frequent itemsets are counted to find the actual frequent itemsets.

Existing parallel mining techniques (Agrawal and Shafer 1999; Chattratichat et al. 1997; Cheung et al. 1996) could also be used to mine different extreme association rules in multi-databases. Zeng et al. (2012) surveyed state-of-the-art algorithms and applications in distributed data mining. Zhong et al. (2003) have proposed a theoretical framework for peculiarity oriented mining in multiple data sources. Zhang et al. (2009) have proposed a nonlinear method, named KEMGP, which adopts kernel estimation method for synthesizing global patterns from local patterns. Shang et al. (2008) have proposed an extension to Piatetsky-Shapiro's minimum interestingness condition to mine association rules in multiple databases.

Yi and Zhang (2007) have proposed a privacy-preserving distributed association rule mining protocol based on a semi-trusted mixer model. Rozenberg and Gudes (2006) have presented their work on association rule mining from distributed vertically partitioned data with the goal of preserving the confidentiality of each database. The authors have presented two algorithms for discovering frequent itemsets and for calculating the confidence of the rules.

Zhu and Wu (2007) have proposed a framework DRAMA for discovering patterns from different databases with patterns' relationships satisfying the user-specified constraints. It builds a HFP-tree from multiple databases, and mine patterns from the HFP-tree by integrating users' constraints into the pattern mining process. Wu et al. (2013) have studied the problem of frequent pattern mining without user-specified gap constraints and proposed PMBC to solve the problem of finding patterns without involving user-specified gap requirements. Given a sequence and a support threshold value, PMBC intends to discover all subsequences with their support values equal to or greater than the given threshold value.

Liu et al. (2010) have presented a top-down mining of sequential patterns (TD-Seq) for mining sequential patterns from high-dimensional stock sequence databases. A two-phase mining method has been proposed, in which a top-down transposition-based searching strategy as well as a new support counting method are exploited.

2.6 Synthesizing an Association Rule

The technique of synthesizing heavy association rules is suitable for the real databases, where the trend of the customers' behavior exhibited in one database is usually present in other databases. In particular, a frequent itemset in one database is usually present in some transactions of other databases even if it does not get extracted. Our estimation procedure captures such trend and estimates the support of a missing association rule in a database. Let $E_1(r, DB)$ be the amount of error in estimating support of a missing association rule r in database DB. Also, let $E_2(r, DB)$ be the level of error in assuming support as 0 for the missing association rule in DB. Then the value of $E_1(r, DB)$ is usually lower than $E_2(r, DB)$. The estimated support and confidence of a missing association rule usually reduce the error of synthesizing heavy association rules in different databases. We would like to estimate the support and confidence of a missing association rule rather assuming it as absent in a database. If an association rule fails to get extracted from database DB, then we assume that DB contributes some amount of support and confidence for the association rule. The support and confidence of an association rule r in database DB satisfy the following inequality:

$$0 \leq supp_a(r, DB) \leq conf_a(r, DB) \leq 1 \tag{2.1}$$

At a given $\alpha = \alpha_0$, we observe that the confidence of an association rule r varies over the interval $[\alpha_0, 1]$ as explained in Example 2.1.

Example 2.1 Let $\alpha = 0.333$. Assume that database D_1 contains the following transactions: $\{a1, b1, c1\}, \{a1, b1, c1\}, \{b2, c2\}, \{a2, b3, c3\}, \{a3, b4\}$ and $\{c4\}$. The support and confidence of association rule r: $\{a1\} \rightarrow \{b1\}$ in D_1 are 0.333 and 1.0 (highest), respectively. Assume that database D_2 contains the following transactions: $\{a1, b1, c1\}, \{a1, b1\}, \{a1, c1\}, \{a1\}, \{a1, b2\}$ and $\{a1, b3\}$. The support and confidence of r in D_2 are 0.333 and 0.333 (lowest), respectively. •

As the support of an association rule is expressed as the lower bound of its confidence, the confidence goes up as support increases. The support of an association rule is distributed over $[0, 1]$. If an association rule is not extracted from a database, then the support falls in $[0, \alpha)$, since the suggested association rules are also considered for synthesizing association rules. We would be interested in estimating the support of such rules. Assume that the association rule r: $\{c\} \rightarrow \{d\}$ has been extracted from m databases, $1 \leq m \leq n$. Without any loss of generality, we assume that the association rule r has been reported from the first m databases. We shall use the average behavior of the customers of the first m branches to estimate the average behavior of the customers in remaining branches. Let $D_{i,j}$ denote the union of databases $D_i, D_{i+1}, \ldots, D_j$, for $1 \leq i \leq j \leq n$. Then, $supp_a(\{c, d\}, D_{1,m})$ could be viewed as the average behavior of customers of the first m branches for purchasing items c and d together at the same time. Then, $supp_a(\{c, d\}, D_{1,m})$ is obtained by using the following formula:

$$supp_a\left(\{c,d\}, D_{1,m}\right) = \left(\sum_{i=1}^{m} supp_a(r, D_i) \times size(D_i)\right) \bigg/ \sum_{i=1}^{m} size(D_i) \quad (2.2)$$

We estimate the support of association rule r for each of the remaining $(n - m)$ databases as follows:

$$supp_s(r, D_{m+1,n}) = \alpha \times supp_a\left(\{c,d\}, D_{1,m}\right) \quad (2.3)$$

The number of the transactions containing the itemset $\{c, d\}$ in D_i is $supp_a(r, D_i) \times size(D_i)$, for $i = 1, 2,..., m$. The association rule r is not present in D_i, for $i = m + 1, m + 2,..., n$. Then the estimated number of the transactions containing the itemset $\{c, d\}$ in D_i is $supp_s(r, D_{m+1, n}) \times size(D_i)$, for $i = m + 1, m + 2,..., n$. The estimated support of association rule r in D_i is determined in the form:

$$supp_e(r, D_i) = \begin{cases} supp_a(r, D_i), & \text{for } i = 1, 2,..., m \\ supp_s(r, D_{m+1,n}), & \text{for } i = m + 1, m + 2,..., n \end{cases} \quad (2.4)$$

Then the synthesized support of association rule r in D is expressed as follows.

$$supp_s(r, D) = \left(\sum_{i=1}^{n} supp_e(r, D_i) \times size(D_i)\right) \bigg/ \sum_{i=1}^{n} size(D_i) \quad (2.5)$$

The confidence of the association rule r depends on the supports of the itemsets $\{c\}$ and $\{c, d\}$. The support of itemset $\{c, d\}$ has been synthesized. Now, we need to synthesize the support of itemset $\{c\}$. Without any loss of generality, let the itemset $\{c\}$ gets extracted from first p databases, for $1 \leq m \leq p \leq n$. The estimated support of frequent itemset $\{c\}$ in D_i is calculated as follows:

$$supp_e(\{c\}, D_i) = \begin{cases} supp_a(\{c\}, D_i), & \text{for } i = 1, 2,..., p \\ supp_s(\{c\}, D_{p+1,n}), & \text{for } i = p + 1, p + 2,..., n \end{cases} \quad (2.6)$$

Then the synthesized support of itemset $\{c\}$ in D is determined.

$$supp_s(\{c\}, D) = \left(\sum_{i=1}^{n} supp_e(\{c\}, D_i) \times size(D_i)\right) \bigg/ \sum_{i=1}^{n} size(D_i) \quad (2.7)$$

We compute the synthesized confidence of association rule r in D.

$$conf_s(r, D) = supp_s(r, D)/supp_s(\{c\}, D) \quad (2.8)$$

2.6.1 Design of Algorithm

Here we present an algorithm for synthesizing heavy association rules in D. The algorithm also indicates whether a heavy association rule is high-frequency rule or exceptional rule. Let N and M be the number of association rules and the number of suggested association rules in different local databases, respectively. The association rules and suggested association rules are kept in arrays RB and SB, respectively. An association rule could be described by following attributes: *ant*, *con*, *did*, *supp* and *conf*. The attributes *ant*, *con*, *did*, *supp* and *conf* represent antecedent, consequent, database identification, support, and confidence of a rule, respectively. An attribute x of the i-th association rule of RB is denoted by $RB(i).x$, $i = 1, 2,\ldots, |RB|$. All the synthesized association rules are kept in array SR. Each synthesized association rule could be described by following attributes: *ant*, *con*, *did*, *ssupp* and *sconf*. The attributes *ssupp* and *sconf* represent synthesized support and synthesized confidence of a synthesized association rule, respectively. In the context of mining heavy association rules in D, the following additional attributes are also considered: *heavy*, *highFreq*, *lowFreq* and *except*. The attributes *heavy*, *highFreq*, *lowFreq* and *except* are used to indicate whether an association rule is a heavy rule, high-frequency rule, low-frequency rule and exceptional rule in D, respectively. An attribute y of the i-th synthesized association rule of SR is denoted by $SR(i).y$, $i = 1, 2,\ldots, |SR|$.

Algorithm 2.1 Synthesize heavy association rules in D. Also, it indicates whether a heavy association rule is high-frequency rule or exceptional rule.

procedure Association-Rule-Synthesis (n, RB, SB, μ, v, size, $\gamma1$, $\gamma2$)

Inputs:
n: number of databases
RB: array of association rules
SB: array of suggested association rules
μ: threshold of high-support for determining heavy association rules
v: threshold of high-confidence for determining heavy association rules
size: array of the number of transactions in different databases
γ_1: threshold of low-frequency for determining low-frequency association rules
γ_2: threshold of high-frequency for determining high-frequency association rules
Outputs:
Heavy association rules along with their high-frequency and exceptionality statuses
01: copy rules of *RB* and *SB* into array *R*;
02: sort rules of *R* based on attributes *ant* and *con*;
03: calculate total number of transactions in all the databases and store it in *totalTrans*;
04: **let** *nSynRules* = 1;
05: **let** *curPos* = 1;
06: **while** (*curPos* $\leq |R|$) **do**
07:   calculate number of occurrences of current rule *R*(*curPos*) and store it in *nExtractions*;
08:   **let** SR(*nSynRules*).*highFreq* = false;
09:   **if** ((*nExtractions* / *n*) $\geq \gamma_2$) **then**
10:     *SR*(*nSynRules*). *highFreq* = true;
11:   **end if**
12:   **let** *SR*(*nSynRules*).*lowFreq* = false;
13:   **if** ((*nExtractions* / *n*) $< \gamma_1$) **then**
14:     *SR*(*nSynRules*).*lowFreq* = true;
15:   **end if**

16:    calculate $supp_s(R(curPos), D)$ using formula (2.5);
17:    calculate $conf_s(R(curPos), D)$ using formula (2.8);
18:    **let** $SR(nSynRules).heavy$ = false;
19:    **if** $((supp_s(SR(nSynRules), D) \geq \mu)$ **and** $(conf_s(SR(nSynRules), D) \geq \nu))$ **then**
20:        $SR(nSynRules).heavy$ = true;
21:    **end if**
22:    **let** $SR(nSynRules).except$ = false;
23:    **if** $((SR(nSynRules)$ is a low-frequency rule) **and** $(SR(nSynRules)$ is a heavy rule)) **then**
24:        $SR(nSynRules).except$ = true;
25:    **end if**
26:    update index $curPos$ for processing the next association rule;
27:    increase index $nSynRules$ by 1;
28: **end while**
29: **for** each synthesized association rule τ in SR **do**
30:    **if** τ is heavy **then**
31:        display τ along with its high-frequency and exceptionality statuses;
32:    **end if**
33: **end for**
end procedure

The above algorithm works as follows. The association rules and suggested association rules are copied into R. All the association rules in R are sorted on the pair of attributes {ant, con}, so that the same association rule extracted from different databases remains together after sorting. Thus, it would help synthesizing a single association rule at a time. The synthesis process is realized in the while-loop shown in line 6. Based on the number of extractions of an association rule, we could determine its high-frequency and low-frequency statuses. The number of extractions of current association rule has been determined as indicated in line 7. The high-frequency status of current association rule is determined—see lines 8–11. Also, the low-frequency status of current association rule is determined (lines 12–15). We synthesize support and confidence of current association rule based on Eqs. (2.5) and (2.8), respectively. Once the synthesized support and synthesized confidence have been determined, we could identify the heavy and exceptional statues of current association rule. The heavy status of current association rule is determined using the part of the procedure covered in lines 18–21. Also, the exceptional status of current association rule is determined using lines 22–25. At line 26, we determine the next association rule in R for the synthesizing process. Heavy association rules are displayed along with their high-frequency and exceptionality statuses using lines 29–33. The shaded lines of the pseudo code have been added to report the high-frequency and exceptional statuses of heavy association rules.

Theorem 2.1 *The time complexity of procedure Association-Rule-Synthesis is maximum$\{O((M + N) \times \log(M + N)), O(n \times (M + N))\}$, where N and M are the number of association rules and the number of suggested association rules extracted from n databases.*

Proof The lines 1 and 2 take time in $O(M + N)$ and $O((M + N) \times \log(M + N))$ respectively, since there are $M + N$ rules in different local databases. The while-loop at line 6 repeats maximum $M + N$ times. Line 7 takes $O(n)$ time, since each rule is extracted maximum n number of times. Lines 8–15 take $O(1)$ time. Using formula (2.3), we could calculate the average behavior of customers of the first m databases in $O(m)$ time. Each of lines 16 and 17 takes $O(n)$ time. Lines 18–25 take $O(1)$ time. Line 26 could be executed during execution of line 7. Thus, the time complexity of while-loop 6–28 is $O(n \times (M + N))$. The time complexity of lines 29–33 is $O(M + N)$, since the number of synthesized association rules is less than or equal to $M + N$. Thus, time complexity of procedure *Association-Rule-Synthesis* is $maximum\{O((M + N) \times \log(M + N)), O(n \times (M + N)), O(M + N)\}$ $= maximum\{O((M + N) \times \log(M + N)), O(n \times (M + N))\}$. •

Wu and Zhang (2003) have proposed *RuleSynthesizing* algorithm for synthesizing high-frequency association rules in different databases. The algorithm is based on the weights of the different databases. Again, the weight of a database would depend on the association rules extracted from the database. The proposed algorithm executes in $O(n^4 \times maxNosRules \times totalRules^2)$ time, where n, *max-NosRules*, and *totalRules* are the number of data sources, the maximum among the numbers of association rules extracted from different databases, and the total number of association rules in different databases, respectively. Ramkumar and Srinivasan (2008) have proposed a modification of *RuleSynthesizing* algorithm. In this modified algorithm, the weight of an association rule is based on the size of a database. This assumption seems to be more logical. For synthesizing confidence of an association rule, the authors have described a method which was originally proposed by Adhikari and Rao (2008). Though the time complexity of modified *RuleSynthesizing* algorithm is the same as that of the original *RuleSynthesizing* algorithm, but it reduces the average error in synthesizing an association rule. The algorithm *Association-Rule-Synthesis* could synthesize heavy association rules, high-frequency association rules, and exceptional association rules in *maximum* $\{O(totalRules \times \log(totalRules)), O(n \times totalRules)\}$ time. Thus, algorithm *Association-Rule-Synthesis* takes much less time than the existing algorithms. Moreover, the proposed algorithm is simple and straight forward. We illustrate the performance of the proposed algorithm using the following example.

Example 2.2 Let D_1, D_2 and D_3 be three databases of sizes 4,000 transactions, 3,290 transactions, and 10,200 transactions, respectively. Let D be the union of the databases D_1, D_2, and D_3. Assume that $\alpha = 0.2$, $\beta = 0.3$, $\gamma_1 = 0.4$, $\gamma_2 = 0.7$, $\mu = 0.3$ and $v = 0.4$. The following association rules have been extracted from the given databases. r_1: $\{H\} \rightarrow \{C, G\}$, r_2: $\{C\} \rightarrow \{G\}$, r_3: $\{G\} \rightarrow \{F\}$, r_4: $\{H\} \rightarrow \{E\}$, r_5: $\{A\} \rightarrow \{B\}$. The rulebases are given as follows: $RB_1 = \{r_1, r_2\}$, $SB_1 = \{r_3\}$; $RB_2 = \{r_4\}$, $SB_2 = \{r_1\}$; $RB_3 = \{r_1, r_5\}$, $SB_3 = \{r_2\}$. The supports and confidences of the association rules are given as follows. $supp_a(r_1, D_1) = 0.22$, $conf_a(r_1, D_1) = 0.55$; $supp_a(r_1, D_2) = 0.25$, $conf_a(r_1, D_2) = 0.29$; $supp_a(r_1, D_3) = 0.20$, $conf_a(r_1, D_3) = 0.52$; $supp_a(r_2, D_1) = 0.69$, $conf_a(r_2, D_1) = 0.82$; $supp_a(r_2, D_3) = 0.23$, $conf_a(r_2, D_3) = 0.28$; $supp_a(r_3, D_1) = 0.22$, $conf_a(r_3,$

Table 2.1 Heavy association rules in the union of databases given in Example 2.2

r: ant → con	Ant	Con	$Supp_s(r, D)$	$Conf_s(r, D)$	Heavy	High frequency	Except
r_2	C	G	0.31	0.66	True	False	False
r_5	A	B	0.57	0.90	True	False	True

$D_1) = 0.29$; $supp_a(r_4, D_2) = 0.40$, $conf_a(r_4, D_2) = 0.45$; $supp_a(r_5, D_3) = 0.86$, $conf_a(r_5, D_3) = 0.92$. Also, let $supp_a(\{A\}, D_3) = 0.90$, $supp_a(\{C\}, D_1) = 0.80$, $supp_a(\{C\}, D_3) = 0.40$, $supp_a(\{G\}, D_1) = 0.29$, $supp_a(\{H\}, D_1) = 0.31$, $supp_a(\{H\}, D_2) = 0.33$, and $supp_a(\{H\}, D_3) = 0.50$. Heavy association rules are presented in Table 2.1.

The association rules r_2 and r_5 have synthesized support greater than or equal to 0.3 and synthesized confidence greater than or equal to 0.4. So, r_2 and r_5 are heavy association rules in D. The association rule r_5 is a exceptional rule, since it is a heavy and low-frequency rule. But the association rule r_2 is neither a high-frequency nor exceptional rule. Though the association rule r_1 is a high-frequency rule but it is not a heavy rule, since $supp_s(r_1, D) = 0.21$ and $conf_s(r_1, D) = 0.48$. •

2.6.2 Error Calculation

To evaluate the proposed technique of synthesizing heavy association rules we have determined the error which has occurred in the experiments. More specifically, the error is expressed relative to the number of transactions, number of items, and the length of a transaction in the databases. Thus the error of an experiment needs to be expressed along with *ANT*, *ALT*, and *ANI* in the given databases, where *ANT*, *ALT* and *ANI* denote the average number of transactions, the average length of a transaction and the average number of items in a database, respectively. There are several ways one could define the error. The proposed definition of error is based on the frequent itemsets generated from heavy association rules. Let $r: \{c\} \to \{d\}$ be a heavy association rule. Then the frequent itemsets generated from association rule r are $\{c\}$, $\{d\}$, and $\{c, d\}$. Let $\{X_1, X_2,..., X_m\}$ be set of frequent itemsets generated from all the heavy association rules in D. We define the following two types of error.

1. *Average Error* (AE)

$$AE(D, \alpha, \mu, v) = \frac{1}{m} \sum_{i=1}^{m} |supp_a(X_i, D) - supp_s(X_i, D)| \tag{2.9}$$

2. *Maximum Error* (ME)

$$ME(D, \alpha, \mu, v) = maximum\{ |supp_a(X_i, D) - supp_s(X_i, D)|, \quad i = 1, 2,..., m \} \tag{2.10}$$

where $supp_a(X_i, D)$ and $supp_s(X_i, D)$ are actual support i.e., the support based on apriori algorithm and synthesized support of the itemset X_i in D, respectively. In Example 2.3, we illustrate the behavior of the measures given above.

Example 2.3 With reference to Example 2.2, r_2: $C \rightarrow G$ and r_5: $A \rightarrow B$ are heavy association rules in D. The frequent itemsets generated between r_2 and r_5 are A, B, C, G, AB and CG. For the purpose of finding the error of an experiment, we need to find the actual supports of the itemsets generated from the heavy association rules. The actual support of an itemset generated from a heavy association rule could be obtained by mining all the databases D_1, D_2, and D_3 together.

Thus,

$$AE(D, 0.2, 0.3, 0.4) = \frac{1}{6}\{|supp_a(\{A\}, D) - supp_s(\{A\}, D)| + |supp_a(\{B\}, D) - supp_s(\{B\}, D)|$$
$$+ |supp_a(\{C\}, D) - supp_s(\{C\}, D)| + |supp_a(\{G\}, D) - supp_s(\{G\}, D)|$$
$$+ |supp_a(\{A, B\}, D) - supp_s(\{A, B\}, D)| + |supp_a(\{C, G\}, D) - supp_s(\{C, G\}, D)|\}.$$

$$ME(D, 0.2, 0.3, 0.4) = maximum\{|supp_a(\{A\}, D) - supp_s(\{A\}, D)|, |supp_a(\{B\}, D) - supp_s(\{B\}, D)|$$
$$|supp_a(\{C\}, D) - supp_s(\{C\}, D)|, |supp_a(\{G\}, D) - supp_s(\{G\}, D)|,$$
$$|supp_a(\{A, B\}, D) - supp_s(\{A, B\}, D)|, |supp_a(\{C, G\}, D) - supp_s(\{C, G\}, D)|\}. \blacksquare$$

2.7 Experiments

We have carried out several experiments to study the effectiveness of the approach presented in this chapter. We present the experimental results using three real databases. The database *retail* (Frequent itemset mining dataset repository 2004) is obtained from an anonymous Belgian retail supermarket store. The databases *BMS-Web-Wiew-1* and *BMS-Web-Wiew-2* can be found from KDD CUP 2000 (Frequent itemset mining dataset repository 2004). We present some characteristics of these databases in Table 2.2. We use notation *DB*, *NT*, *AFI*, *ALT* and *NI* to denote a database, the number of transactions, the average frequency of an item, the average length of a transaction and the number of items in the database, respectively.

Each of the above databases is divided into 10 subsets for the purpose of carrying out experiments. The databases obtained from *retail*, *BMS-Web-Wiew-1* and *BMS-Web-Wiew-2* are named as R_i, B_{1i} and B_{2i} respectively, $i = 0, 1,..., 9$.

Table 2.2 Dataset characteristics

Dataset	NT	ALT	AFI	NI
Retail	88,162	11.31	99.67	10,000
BMS-Web-Wiew-1	1,49,639	2.00	155.71	1,922
BMS-Web-Wiew-2	3,58,278	2.00	7165.56	100

Table 2.3 Branch database characteristics

DB	NT	ALT	AFI	NI	DB	NT	ALT	AFI	NI
R_0	9,000	11.24	12.07	8,384	R_5	9,000	10.86	16.71	5,847
R_1	9,000	11.21	12.27	8,225	R_6	9,000	11.20	17.42	5,788
R_2	9,000	11.34	14.60	6,990	R_7	9,000	11.16	17.35	5,788
R_3	9,000	11.49	16.66	6,206	R_8	9,000	12.00	18.69	5,777
R_4	9,000	10.96	16.04	6,148	R_9	7,162	11.69	15.35	5,456
B_{10}	14,000	2.00	14.94	1,874	B_{15}	14,000	2.00	280.00	100
B_{11}	14,000	2.00	280.00	100	B_{16}	14,000	2.00	280.00	100
B_{12}	14,000	2.00	280.00	100	B_{17}	14,000	2.00	280.00	100
B_{13}	14,000	2.00	280.00	100	B_{18}	14,000	2.00	280.00	100
B_{14}	14,000	2.00	280.00	100	B_{19}	23,639	2.00	472.78	100
B_{20}	35,827	2.00	1326.93	54	B_{25}	35,827	2.00	716.54	100
B_{21}	35,827	2.00	1326.93	54	B_{26}	35,827	2.00	716.54	100
B_{22}	35,827	2.00	716.54	100	B_{27}	35,827	2.00	716.54	100
B_{23}	35,827	2.00	716.54	100	B_{28}	35,827	2.00	716.54	100
B_{24}	35,827	2.00	716.54	100	B_{29}	35,835	2.00	716.70	100

The databases R_j and B_{ij} are called branch databases, $i = 1, 2$, and $j = 0, 1,..., 9$. Some characteristics of the branch databases are presented in Table 2.3.

The results of the three experiments using Algorithm 2.1 are presented in Table 2.4. The choice of different parameters is an important issue. We have selected different values of α and β for different databases. But, they are kept the same for branch databases obtained from the same database. For example, α and β are the same for branch databases R_i, for $i = 0, 1,..., 9$.

Table 2.4 First five heavy association rules reported from different databases (sorted in non-increasing order on synthesized support)

Data base	α	β	μ	ν	Heavy assoc rules	Syn supp	Syn conf	High freq	Exceptional
$\cup_{i=0}^{9} R_i$	0.05	0.2	0.1	0.5	$\{48\} \rightarrow \{39\}$	0.33	0.68	Yes	No
					$\{39\} \rightarrow \{48\}$	0.33	0.56	Yes	No
					$\{41\} \rightarrow \{39\}$	0.13	0.63	Yes	No
					$\{38\} \rightarrow \{39\}$	0.12	0.66	Yes	No
					$\{41\} \rightarrow \{48\}$	0.10	0.51	Yes	No
$\cup_{i=0}^{9} B_{1i}$	0.01	0.2	0.007	0.1	$\{1\} \rightarrow \{5\}$	0.01	0.13	No	No
					$\{5\} \rightarrow \{1\}$	0.01	0.11	No	No
					$\{7\} \rightarrow \{5\}$	0.01	0.12	No	No
					$\{5\} \rightarrow \{7\}$	0.01	0.11	No	No
					$\{3\} \rightarrow \{5\}$	0.01	0.12	No	No
$\cup_{i=0}^{9} B_{2i}$	0.006	0.01	0.01	0.1	$\{3\} \rightarrow \{1\}$	0.02	0.14	Yes	No
					$\{1\} \rightarrow \{3\}$	0.02	0.14	Yes	No
					$\{7\} \rightarrow \{1\}$	0.02	0.14	Yes	No
					$\{1\} \rightarrow \{7\}$	0.02	0.14	Yes	No
					$\{5\} \rightarrow \{1\}$	0.02	0.14	Yes	No

2.7.1 Results of Experimental Studies

After mining a branch database from a group of branch databases using a reasonably low values α and β, one could fix α and β for the purpose data mining task. If α and β are smaller, then synthesized support and synthesized confidence values are closer to their actual values. Thus, the synthesized association rules are closer to the true association rules in multiple databases.

The choice of the values of μ and v are context dependent. Also if μ and v are kept fixed then some databases might not report heavy association rules, while other databases might report many heavy association rules. While generating association rule one could estimate the average synthesized support and confidence based on the generated association rules. Thus, it gives an idea of thresholds for high-support and high-confidence for synthesizing heavy association rules in different databases. Also, the choice of γ_1 and γ_2 are also context dependent. It has been found that "reasonable" values of γ_1 and γ_2 could lie in the interval [0.3, 0.4] and [0.6, 0.7], respectively. Given these findings, we have taken $\gamma_1 = 0.35$, and $\gamma_2 = 0.60$ for synthesizing heavy association rules.

The experiments conducted on the three databases have resulted in no exceptional association rule. Normally, exceptional association rules are rare. Also, we have not found any association rule which is a heavy rule as well as high-frequency rule in multiple databases obtained from *BMS-Web-Wiew-1*.

In many applications, the suggested association rules are significant. While synthesizing the association rules from different databases we might need to consider the suggested association rules for the correctness of synthesizing association rules. We have observed that the number of suggested association rules in the set of databases $\{R_0, R_1,..., R_9\}$ and $\{B_{10}, B_{11},..., B_{19}\}$ are significant. But, the set of databases $\{B_{20}, B_{21},..., B_{29}\}$ do not generate any suggested association rule. We present the number of association rules and the number of suggested association rules for different experiments in Table 2.5.

The error of synthesizing association rules in a database is relative to the following parameters: the number of transactions, the number of items, and the length of transactions in the given databases. If the number of transactions in database increases, the error of synthesizing association rules also increases, provided other two parameters remain constant. If the lengths of transactions of a database increase, the error of synthesizing association rules is likely to increase, provided that two other parameters remain constant. Lastly, if the number of items

Table 2.5 Number of association rules and suggested association rules extracted from multiple databases

Database	α	β	Number of association rules (N)	Number of suggested association rules (M)	$M/(N+M)$
$\cup_{i=0}^{9} R_i$	0.05	0.2	821	519	0.39
$\cup_{i=0}^{9} B_{1i}$	0.01	0.2	50	96	0.66
$\cup_{i=0}^{9} B_{2i}$	0.006	0.01	792	0	0

Table 2.6 Error of synthesizing different extreme association rules

Database	α	β	μ	v	(AE, ANT, ALT, ANI)	(ME, ANT, ALT, ANI)
$\cup_{i=0}^{9} R_i$	0.05	0.2	0.1	0.5	(0.00, 8816.2, 11.31, 5882.1)	(0.00, 8816.2, 11.31, 5882.1)
$\cup_{i=0}^{9} B_{1i}$	0.01	0.2	0.007	0.1	(0.00, 14963.9, 2.0, 277.4)	(0.00, 14963.9, 2.0, 277.4)
$\cup_{i=0}^{9} B_{2i}$	0.006	0.01	0.01	0.1	(0.000118, 35827.8, 2.0, 90.8)	(0.00, 35827.8, 2.0, 90.8)

increases, then the error of synthesizing association rules is likely to decrease, provided that two other parameters remain constant. Thus, the error needs to be reported along with the *ANT*, *ALT* and *ANI* for the given databases. The obtained results are presented in Table 2.6.

2.7.2 Comparison with Existing Algorithm

In this section, we make a detailed comparison among the part of the proposed algorithm that synthesizes only high-frequency association rules, *RuleSynthesizing* algorithm (Wu and Zhang 2003) and *Modified RuleSynthesizing* algorithm (Ramkumar and Srivinasan 2008). Let the part of the proposed algorithm be *High-Frequency-Rule-Synthesis* used for synthesizing (only) high-frequency association rules in different databases. We conduct experiments for comparing these algorithms. We compare them on the basis of the following two criteria, namely average error and execution time.

2.7.2.1 Analysis of Average Error

The definitions of average error and maximum error given above and those proposed by Wu and Zhang (2003) are similar and use the same set of synthesized frequent itemsets. However the methods of synthesizing frequent itemsets for these two approaches are different. Thus, the value of error incurred in these two approaches might differ. In *RuleSynthesizing* algorithm, if an itemset fails to get extracted from a database then the support of the itemset is assumed to be 0. But in *Association-Rule-Synthesis* algorithm, if an itemset fails to get extracted from a database then the support of the itemset is estimated. The synthesized support of an itemset in the union of databases in these two approaches might be different. As the number of databases increases the relative presence of a rule normally decreases. The error of synthesizing an association rule normally increases. The AE reported in the experiment is likely to increase if the number of databases increases. We observe such phenomenon in Figs. 2.2 and 2.3.

The proposed algorithm follows a direct approach in identifying high-frequency association rules as opposed to the *RuleSynthesizing* and *Modified RuleSynthesizing* algorithms. In Figs. 2.2 and 2.3, we observe that AE of an experiment conducted using *High-Frequency-Rule-Synthesis* algorithm is less than that of

Fig. 2.2 AE versus the number of databases from *retail* at $(\alpha, \beta, \gamma) = (0.05, 0.2, 0.6)$

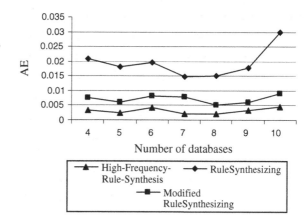

Fig. 2.3 AE versus the number of databases from *BMS-Web-Wiew-1* at $(\alpha, \beta, \gamma) = (0.005, 0.1, 0.3)$

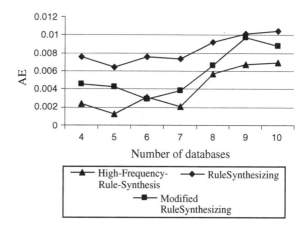

RuleSynthesizing algorithm. But the *Modified RuleSynthesizing* algorithm improves the accuracy of synthesizing an association rule as compared to *Rule-Synthesizing* algorithm. It remains less accurate when compared to the accuracy of the *High-Frequency-Rule-Synthesis* algorithm.

2.7.2.2 Analysis of Execution Time

We have also completed experiments to study the execution time by varying the number of databases. The number of synthesized frequent itemsets increases as the number of databases increases. The execution time increases with the increase of number of databases. We observe this phenomenon in Figs. 2.4 and 2.5. However, more significant differences are noted with the increase in the number of databases.

The time complexities of *RuleSynthesizing* and *Modified RuleSynthesizing* algorithms is the same. When the number of databases are less the *RuleSynthesizing*

Fig. 2.4 Execution time versus the number of databases from *retail* at $(\alpha, \beta, \gamma) = (0.05, 0.2, 0.6)$

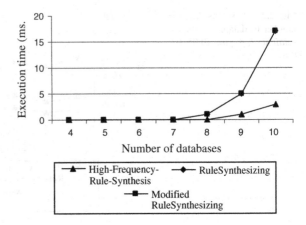

Fig. 2.5 Execution time versus the number of databases from *BMS-Web-Wiew-1* at $(\alpha, \beta, \gamma) = (0.005, 0.1, 0.3)$

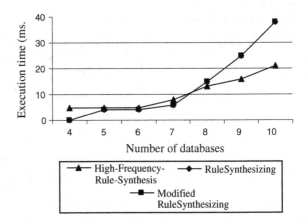

and *Modified RuleSynthesizing* algorithms might be faster than *High-Frequency-Rule-Synthesizing* algorithm. As the number of databases increases, *High-Frequency-Rule-Synthesizing* algorithm works faster than both *RuleSynthesizing* and *Modified RuleSynthesizing* algorithms.

2.8 Conclusions

The extended model of local pattern analysis enables us to develop useful multi-database mining applications. Although it exhibits many layers and interfaces, this general model can come with many variations. In particular, some of these layers might not be present when developing a certain application. Synthesizing heavy association rule is an important component of a multi-database mining system. In this chapter, we have presented three extreme types of association rules present in multiple databases viz., heavy association rules, high-frequency association rules, and exceptional association rules. The introduced algorithm referred to as the *Association-Rule-Synthesis* is used to synthesize these extreme association rules in multiple databases.

References

Adhikari A, Rao PR (2008) Synthesizing heavy association rules from different real data sources. Pattern Recogn Lett 29(1):59–71

Agrawal R, Shafer J (1999) Parallel mining of association rules. IEEE Trans Knowl Data Eng 8(6):962–969

Agrawal R, Srikant R (1994) Fast algorithms for mining association rules. In: Proceedings of international conference on very large data bases, pp 487–499

Agrawal R, Imielinski T, Swami A (1993) Mining association rules between sets of items in large databases. In: Proceedings of ACM SIGMOD conference, pp 207–216

Chattratichat J, Darlington J, Ghanem M, Guo Y, Hüning H, Köhler M, Sutiwaraphun J, To HW, Yang D (1997) Large scale data mining: Challenges, and responses. In: Proceedings of the third international conference on knowledge discovery and data mining, pp 143–146

Cheung D, Ng V, Fu A, Fu Y (1996) Efficient mining of association rules in distributed databases. IEEE Trans Knowl Data Eng 8(6):911–922

Frequent itemset mining dataset repository (2004) http://fimi.cs.helsinki.fi/data

Han J, Pei J, Yiwen Y (2000) Mining frequent patterns without candidate generation. In: Proceedings of ACM SIGMOD conference on management of data, pp 1–12

Last M, Kandel A (2001) Automated detection of outliers in real-world data. In: Proceedings of the second international conference on intelligent technologies, pp 292–301

Liu H, Lin F, He J, Cai Y (2010) New approach for the sequential pattern mining of high-dimensional sequence databases. Decis Support Syst 50(1):270–280

Pyle D (1999) Data preparation for data mining. Morgan Kufmann, San Francisco

Ramkumar T, Srivinasan R (2008) Modified algorithms for synthesizing high-frequency rules from different data sources. Knowl Inf Syst 17(3):313–334

Rozenberg B, Gudes E (2006) Association rules mining in vertically partitioned databases. Data Knowl Eng 59(2):378–396

Savasere A, Omiecinski E, Navathe S (1995) An efficient algorithm for mining association rules in large databases. In: Proceedings of the 21st international conference on very large data bases, pp 432–443

Shang S, Dong X, Li J, Zhao Y (2008) Mining positive and negative association rules in multi-database based on minimum interestingness. In: Proceedings of the 2008 international conference on intelligent computation technology and automation, pp 791–794

Wu X, Zhang S (2003) Synthesizing high-frequency rules from different data sources. IEEE Trans Knowl Data Eng 14(2):353–367

Wu X, Zhu X, He Y, Abdullah N, Arslan AN (2013) PMBC: Pattern mining from biological sequences with wildcard constraints. Comp Bio Med 43(5):481–492

Yi X, Zhang Y (2007) Privacy-preserving distributed association rule mining via semi-trusted mixer. Data Knowl Eng 63(2):550–567

Zeng L, Li L, Duan L, Lü K, Shi Z, Wang M, Wu W, Luo P (2012) Distributed data mining: a survey. Inf Technol Manage 13(4):403–409

Zhang S, Wu X (2011) Fundamentals of association rules in data mining and knowledge discovery. Wiley Interdisc Rev: Data Min Knowl Discovery 1(2):97–116

Zhang S, Wu X, Zhang C (2003) Multi-database mining. IEEE Comput Intell Bull 2(1):5–13

Zhang S, You X, Jin Z, Wu X (2009) Mining globally interesting patterns from multiple databases using kernel estimation. Expert Syst Appl: An Int J 36(8):10863–10869

Zhong N, Yao YYY, Ohishima M (2003) Peculiarity oriented multidatabase mining. IEEE Trans Knowl Data Eng 15(4):952–960

Zhu X, Wu X (2007) Discovering relational patterns across multiple databases. In: Proceedings of ICDE, pp 726–735

Chapter 3
Clustering Items in Time-Stamped Databases Induced by Stability

Many multi-branch companies transact from different branches. Each branch of such a company maintains a separate database over time. The variation of sales of an item over time is an important issue, and therefore, we present the notion of stability of an item. Stable items are useful in making numerous strategic decisions for a company. We have discussed two measures of stability of an item. Based on the degree of stability of an item, an algorithm is designed for finding partition among items in different data sources. Then the notion of the best cluster is introduced by considering average degree of variation of a class, and designed an algorithm to find clusters among items in different data sources. The best cluster is determined by average degree of variation in a cluster. Experimental results are provided for three transactional databases.

3.1 Introduction

Due to a liberalization of government policies across the globe, a number of multi-branch companies has been steadily increasing over time. There are multi-branch companies that deal with multiple databases. Thus, the study of data mining on multiple databases arises as an important issue. Knowledge discovery in multiple databases has been recently recognized as an important area of research in data mining (Zhang et al. 2004; Adhikari et al. 2010). Many multi-branch companies deal with transactional time-stamped data. Transactional data collected over time at no particular frequency are often termed as transactional time-stamped data (Leonard and Wolfe 2005). Some examples of transactional time-stamped data are point of sales data, inventory data, and trading data. Little work has been reported on the area of mining multiple transactional time-stamped databases. In this chapter, we discuss a useful data mining application on multiple transactional time-stamped data.

Consider a multi-branch company that transacts from multiple branches, and all the transactions in a branch are stored locally. A transaction could be viewed as a collection of items with a unique identifier. Then it becomes interesting to study

A. Adhikari et al., *Data Analysis and Pattern Recognition in Multiple Databases*,
Intelligent Systems Reference Library 61, DOI: 10.1007/978-3-319-03410-2_3,
© Springer International Publishing Switzerland 2014

the characteristics of items in multiple databases. A useful characteristic of item is its variation of sales over time. The items having lower variation of sales over time are useful in devising strategies of the company. Thus it is important to identify such items. Consider a situation as presented in Example 3.1.

Example 3.1 Let there be ten branch databases. Suppose we are interested in studying variations of five items. We take here the support (Agrawal et al. 1993) series of the five items. Let i-th series be the series of supports in ten databases corresponds to item x_i, $i = 1, 2, 3, 4, 5$.

(1) 0.03, 0.20, 0.31, 0.11, 0.07, 0.35, 0.82, 0.62, 0.44, 0.13
(2) 0.19, 0.20, 0.18, 0.21, 0.20, 0.20, 0.19, 0.18, 0.21, 0.20
(3) 0.05, 0.11, 0.07, 0.20, 0.16, 0.12, 0.13, 0.08, 0.17, 0.10
(4) 0.03, 0.04, 0.03, 0.07, 0.08, 0.12, 0.09, 0.15, 0.17, 0.12
(5) 0.04, 0.04, 0.03, 0.05, 0.04, 0.06, 0.04, 0.05, 0.06, 0.05

Among the support series corresponding to different items, it is observed that the variation of sales corresponding to item x_5 is the least. Thus, the strategies based on item x_5 could be deemed more reliable. •

The chapter is organized as follows. Work related to the problem is presented in Sect. 3.2. In Sect. 3.3, we propose a model of mining multiple transactional time-stamped databases. The problem is formulated in Sect. 3.4. In Sect. 3.5, we design an algorithm for clustering of items in multiple databases. Experimental results are provided in Sect. 3.6.

3.2 Related Work

Liu et al. (2001) have proposed stable association rules based on testing of hypothesis. The distribution of test statistic under null hypothesis follows normal distribution for large sample size. Thus, the stable association rules are determined based on some assumptions. Due to this reason, we present a construction of stable items based on the concept of stationary time series data (Brockwell and Richard 2002).

In the context of interestingness measures, Tan et al. (2002) have described several key properties of twenty one interestingness measures proposed in statistics, machine learning and data mining literature. Wu et al. (2005) have proposed two item-based similarity measures for clustering a set of databases. Based on transaction similarity, Adhikari and Rao (2008b) have proposed two similarity measures for clustering databases.

Zhang et al. (1997) have proposed an efficient and scalable data clustering method BIRCH based on a new in-memory data structure called CF-tree. Estivill-Castro and Yang (2004) have proposed an algorithm that remains efficient, generally applicable, multi-dimensional but is more robust to noise and outliers.

Jain et al. (1999) have presented an overview of pattern clustering methods from a statistical pattern recognition perspective, with a goal of providing useful advice and references to fundamental concepts accessible to the broad community of clustering practitioners. In this chapter, we cluster items in multiple databases based on supports of items. Thus, the above algorithms might not be suitable under this framework.

Yang and Shahabi (2005) have proposed an algorithm to determine the stationarity of multivariate time series data for improving the efficiency of many correlation-based data schemes. Matsubara et al. (2012) have proposed *TriMine* algorithm for mining meaningful patterns and forecasting complex time-stamped events. Guil and Marín (2012) have presented a tree-based structure and an algorithm, called *TSET–Miner*, for frequent temporal pattern mining from time-stamped datasets. The algorithm is based on mining inter-transaction association, and is characterized by the use of a single tree-based data structure for generation and storage of all frequent sequences discovered through mining. Albanese et al. (2013) have proposed to start with a known set A of activities (both innocuous and dangerous) that authors wish to monitor. In addition, authors wish to identify "unexplained" subsequences in a sequence of observations that are poorly explained by A (e.g., because they may contain occurrences of activities that have never been seen or anticipated before, i.e. they are not in A). Authors formally defined the probability that a sequence of observations was unexplained totally or partially with respect to (w.r.t.) A, and developed algorithms to identify the top-k totally and partially unexplained sequences w.r.t. A.

In multi-database environment we have proposed the notion of high frequency itemsets, and an algorithm for synthesizing supports of such itemsets is designed (Adhikari 2013). The existing clustering technique might cluster local frequency items at a low level, since it estimates association among items in an itemset with a low accuracy. Therefore, in this work a new algorithm for clustering local frequency items is proposed.

Zhang et al. (2003) designed a local pattern analysis for mining multiple databases. Zhang et al. (2004) studied various issues related to multi-database mining. We have presented some strategies on developing multi-database mining applications (Adhikari et al. 2010).

3.3 A Model of Mining Multiple Transactional Time-Stamped Databases

Consider a multi-branch company that has n branches. All the local transactions are stored locally. Let D_i be the transactional time-stamped database corresponding to i-th branch, $i = 1, 2, ..., n$. Web sites and transactional databases contain a large amount of time-stamped data related to suppliers and/or customers of the organization over time. Mining time-stamped data could help business leaders make better decisions by listening to their suppliers or customers via their

transactions collected over time (Leonard and Wolfe 2005). We present here a
model of mining global patterns in multi-databases over time.

We have proposed an extended model of local pattern analysis for mining
multiple large databases (Adhikari and Rao 2008a). The limitation of this model is
that it may return approximate global patterns. We present here a new model of
mining global patterns in multiple time-stamped databases as shown in Fig. 3.1
(Adhikari et al. 2009). It has a set of interfaces and a set of layers. Each interface is a
set of operations that produces dataset(s) (or, knowledge) based on the lower layer
dataset(s). There are five distinct interfaces of the proposed model of synthesizing
global patterns from local patterns. The function of each interface is described
below. Interface 2/1 cleans/transforms/integrates/reduces data at the lowest layer.
By applying these procedures we get processed database from the original database.
In addition, interface 2/1 applies a filtering algorithm on each database for sepa-
rating relevant data from outlier data. Also, it loads data into the respective data
warehouse. At interface 3/2, each processed database PD_i is partitioned into k time
databases DT_{ij}, where DT_{ij} is the processed database (if available) for the j-th time
slot at the i-th branch, $j = 1, 2, ..., k; i = 1, 2, ..., n$. At interface 4/3 the j-th time
databases of all branches are merged into a single time database DT_j, for $j = 1, 2,$
$..., k$. A traditional data mining technique could be applied on database DT_j at the
interface 5/4, $j = 1, 2, ..., k$. Let PB_j be pattern base corresponding to the time
database $DT_j, j = 1, 2, ..., k$. Finally, all the pattern bases are processed for synthesizing knowledge or, making decision at the interface 6/5. Undirected lines in
Fig. 3.1 are assumed as directed from bottom to top. The proposed model of mining

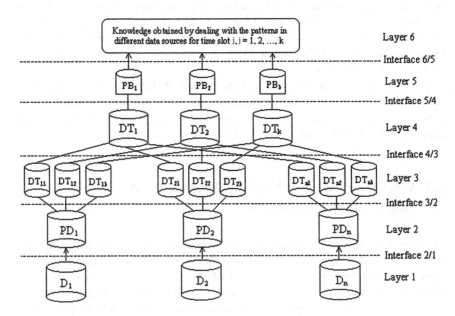

Fig. 3.1 A model of mining global patterns in multiple time-stamped databases

global patterns over time is efficient, since we form the exact global patterns in multiple databases over time.

At layer 4, we have a collection of all time databases. If any one of these databases is too large to apply a traditional data mining technique then this data mining model would fail. In this situation, we could apply an appropriate sampling technique to reduce the size of a database. Thus, we might develop approximate patterns over time.

3.4 Problem Statement

With reference to Fig. 3.1, let DT_j be the database corresponding to the j-th year, $j = 1, 2, ..., k$. Each of these databases corresponds to a specific period of time. Thus, we could call them as time databases. Each of these time databases is mined using a traditional data mining technique (Han et al. 2000; Agrawal and Srikant 1994). For the specific requirement of this problem, we need to mine only items in the time databases. Let I be the set of all items in these databases. Each itemset X in a database D is associated with a statistical measure, called *support* (Agrawal et al. 1993), denoted by $supp(X, D)$. The support of an itemset is defined as the fraction of transactions containing the itemset. The variation of sales of an item over the time is an important aspect in determining stability of the item. Stable items are useful in many applications. Stable items as well could be useful to promote sales of other items. Modelling with stable items is more justified than modelling with unstable items.

Let $\mu_{s(x)}(t)$ be the mean support of item x in the database $DT_1, DT_2, ..., DT_t$. The value of $\mu_{s(x)}(t)$ is obtained by the following formula:

$$\mu_{s(x)}(t) = \frac{\left(\sum_{i=1}^{t} supp(x, DT_i) \times size(DT_i) \right)}{\sum_{i=1}^{t} size(DT_i),} \qquad t = 1, 2, ..., k \qquad (3.1)$$

Let $\sigma(\mu_{s(x)})$ be the standard deviation of $\mu_{s(x)}(t)$, $t = 1, 2, ..., k$. We call $\sigma(\mu_{s(x)})$ the *variation of means* corresponding to support of x. Let $\gamma_{s(x)}(t, t + h)$ be the autocovariance of $supp(x, DT_t)$ at lag h, $t = 1, 2, ..., k - 1$. Thus, $\gamma_{s(x)}(t, t + h)$ is expressed as follows:

$$\gamma_{s(x)}(t, t + h) = \frac{1}{k} \sum_{t=1}^{k-h} \left(supp(x, DT_t) - \mu_{s(x)}(k) \right) \left(supp(x, DT_{t+h}) - \mu_{s(x)}(k) \right)$$

$$(3.2)$$

$\sigma(\gamma_{s(x)}(t, t + h))$ be the standard deviation of $\gamma_{s(x)}(t, t + h)$, $h = 1, 2, ..., k - 1$. We call this $\sigma(\gamma_{s(x)}(t, t + h))$ a *variation of autocovariances* corresponding to support of x. We have chosen standard deviation as a measure of dispersion (Bluman 2006). Standard deviation and mean deviation about mean are relevant measures of dispersion. Unlike the range measure of dispersion these measures take into account a variation due to each support. Skewness, being a descriptive measure of dispersion, is not suitable in this context. Before we define stability of an item, we study the following time series of supports corresponding to an item. In Example 3.2, we compute $\sigma(\mu)$ and $\sigma(\gamma)$ of support series corresponding to different items.

Example 3.2 We continue with Example 3.1. The variations of means and autocovariances of above series are given as follows: (1) $\sigma(\mu) = 0.09342$, $\sigma(\gamma) = 0.01234$, (2) $\sigma(\mu) = 0.00230$, $\sigma(\gamma) = 0.00002$, (3) $\sigma(\mu) = 0.02351$, $\sigma(\gamma) = 0.00039$, (4) $\sigma(\mu) = 0.02114$, $\sigma(\gamma) = 0.00076$, (5) $\sigma(\mu) = 0.0027986$, $\sigma(\gamma) = 0.0000124$. After adding $\sigma(\mu) + \sigma(\gamma)$ for each series, we observe that the value of total variation, $\sigma(\mu) + \sigma(\gamma)$, is the least corresponding to (5), i.e., item x_5. •

We define stable items based on the concept of stationary time series (Brockwell and Richard 2002). In finding $\sigma(\mu)$, we first compute a set of means of support values. Then we compute standard deviation of these mean values. Thus, we find standard deviation of a set of fractions. In finding $\sigma(\gamma)$, we first compute a set of autocovariances of support values. Then we compute standard deviation of these autocovariances. The autocovariance of supports is an average of a set of squared fractions. Thus, we find standard deviation of a set of squared fractions. So, $\sigma(\mu) \geq \sigma(\gamma)$. In fact, $\sigma(\gamma)$ is close to 0. Thus, we define our first measure of stability $stable_1$ (Adhikari et al. 2009) as follows.

Definition 3.1 An item x is stable if $\sigma(\mu_{s(x)}) \leq \delta$, where δ is a user-defined maximum threshold.

More specifically, we may wish to impose restrictions on both $\sigma(\mu)$ and $\sigma(\gamma)$. We define the second measure of stability $stable_2$ (Adhikari et al. 2009) as follows.

Definition 3.2 An item x is stable if $\sigma(\mu_{s(x)}) + \sigma(\gamma_{s(x)}) \leq \delta$, where δ is user-defined maximum threshold.

In Definition 3.2, the expression $\sigma(\mu_{s(x)}) + \sigma(\gamma_{s(x)})$ is the determining factor of stability of an item. We define a degree of variation of an item x as follows

$$degOfVar(x) = \sigma(\mu_{s(x)}) + \sigma(\gamma_{s(x)}). \qquad (3.3)$$

Higher values of *degOfVar* imply lower degrees of stability of the item. Based on above discussion, we state our problem as follows.

Let D_i and DT_j be the databases corresponding to i-th branch and j-th year of a multi-branch company as depicted in Fig. 3.1 respectively, $i = 1, 2, ..., n$; $j = 1, 2, ..., k$. Each of the time (year) databases has been mined using a traditional data mining technique. Based on the mining results, degree of variation of each item has been computed as discussed above. Find the best non-trivial partition (if it exists) of the items in $D_1, D_2, ..., D_n$ based on degree of variation of an item.

A partition (Liu 1985) is linked with a certain type of clustering. A formal definition of a non-trivial partition is given in the following section.

3.5 Clustering Items

The proposed clustering technique is based on the notion of degree of stability of an item. Again, the degree of stability is based on the variations of means and autocovariances. The clustering technique requires computing the degree of variation for each item in the databases. Let I be the set of all items in the databases. Given a set of yearly databases, the difference in variations between every pair of items could be expressed by a square matrix, called *difference in variation (diffInVar)*. We define the difference in variation between items x_i and x_j as follows.

$$diffInVar(i,j) = \left| degOfVar(x_i) - degOfVar(x_j) \right|, \quad for\, x_i, x_j \in I. \quad (3.4)$$

In the following example, we compute below the *diffInVar* matrix corresponding to Example 3.1.

Example 3.3 We continue here with Example 3.1. Matrix *diffInVar* is given as follows

$$diffInVar = \begin{bmatrix} 0 & 0.103 & 0.082 & 0.084 & 0.102 \\ 0.103 & 0 & 0.021 & 0.019 & 0.001 \\ 0.082 & 0.021 & 0 & 0.002 & 0.020 \\ 0.084 & 0.019 & 0.002 & 0 & 0.018 \\ 0.102 & 0.001 & 0.020 & 0.018 & 0 \end{bmatrix}.$$

Matrix *diffInVar* is symmetric. We use this matrix for clustering items in multiple databases. •

Intuitively, if the difference in variations between two items is close to zero then they may be assigned to the same class. We would like to cluster the items based on this idea. Before clustering the items, we define a class as follows.

Definition 3.3 Let $I = \{i_1, i_2, \ldots, i_p\}$ be the set of items. A class formed at the level of difference in variation of α is expressed as follows.

$$class(I, \alpha) = \begin{cases} X{:}X \subseteq I, |X| \geq 2, and\, degOfVar(x_1, x_2) \leq \alpha, & for\, x_1, x_2 \in X \\ X{:}X \subseteq I, |X| = 1 \end{cases}$$

Based on the above definition of a class, we introduce a concept of clustering as follows.

Definition 3.4 Let $I = \{i_1, i_2, \ldots, i_p\}$ be the set of items. Let $\pi(I, \alpha)$ be a clustering of items in I at the level of difference in variation α. Then, $\pi(I, \alpha) = \{X : X \in \rho(I), and\, X\, is\, a\, class\, (I, \alpha)\}$, where $\rho(I)$ is the power set of I.

During the clustering process we may like to impose a restriction that each item belongs to at least one class. This restriction makes a clustering complete. We define a complete clustering as follows.

Definition 3.5 Let $I = \{i_1, i_2, ..., i_p\}$ be the set of items. Let $\pi(I, \alpha) = \{C_1(I, \alpha),$ $C_2(I, \alpha), ..., C_m(I, \alpha)\}$, where $C_k(I, \alpha)$ is the k-th class of the clustering π, for $k = 1, 2, ..., m$. π is complete, if $\bigcup_{k=1}^{m} C_k(I, \alpha) = I$.

In a complete clustering, two classes may have common items. We may be interested in finding out a cluster containing mutually exclusive classes. A mutually exclusive cluster is defined as follows.

Definition 3.6 Let $I = \{i_1, i_2, ..., i_p\}$ be the set of items. Let $\pi(I, \alpha) = \{C_1(I, \alpha),$ $C_2(I, \alpha), ..., C_m(I, \alpha)\}$, where $C_k(I, \alpha)$ is the k-th class of the clustering π, $k = 1, 2, ..., m$. π is mutually exclusive if $C_i(I, \alpha) \cap C_j(I, \alpha) = \phi$, for $i \neq j$, and $1 \leq i, j \leq m$.

We may be interested in finding such a mutually exclusive and complete cluster. A partition of a set of items I is defined as follows.

Definition 3.7 Let $\pi(I, \alpha)$ be a mutually exclusive and complete clustering of a set of items I at the level of difference in variation α. $\pi(I, \alpha)$ is called a non-trivial partition if $1 < |\pi| < m$.

A partition is a cluster. But a cluster is not necessarily a partition. In the next section, we find the best non-trivial partition (if it exists) of a set of items.

The items in a class are similar with respect to their variations. We are interested in the classes of a partition where the variations of items are less. The items in these classes are useful in devising strategies for the company. Thus, we define an average degree of variation *adv*, of a class, as follows.

Definition 3.8 Let C be a class of partition π. Then, $adv(C|\pi) = \frac{1}{|C|}$ $\Sigma_{x \in C} degOfVar(x)$.

3.5.1 Finding the Best Non-Trivial Partition

With reference to Example 3.1, we arrange all non-zero and distinct values of *diffInVar* in non-decreasing order for finding all the non-trivial partitions, $1 \leq i < j \leq 5$. The arranged values of *diffInVar* are given as follows: 0.001, 0.002, 0.018, 0.019, 0.020, 0.021, 0.082, 0.084, 0.102, 0.103. We get two non-trivial partitions at $\alpha = 0.001$, and 0.002. The partitions are given as follows: $\pi^{0.001} = \{\{x_1\}, \{x_2, x_5\}, \{x_3\}, \{x_4\}\}$, and $\pi^{0.002} = \{\{x_1\}, \{x_2, x_5\}, \{x_3, x_4\}\}$. We observe that at different levels of α we come up with different partitions. We would like to determine the best partition among these partitions. The best partition is the one which maximizes the intra-class variation and minimizes the inter-class similarity. Intra-class variation and inter-class similarity are defined as follows.

Definition 3.9 The intra-class variation *intra-var*, of a partition π at the level α is defined as follows: $intra-var(\pi^{\alpha}) = \Sigma_{k=1}^{|\pi|}\Sigma_{x_i,x_j \in C_k; x_i < x_j}|degOfVar(x_i) - degOfVar(x_j)|$

Definition 3.10 The inter-class similarity *inter-sim*, of a partition π at the level α is defined as follows: $inter-sim(\pi^{\alpha}) = \Sigma_{c_p, c_q \in \pi; p < q}\Sigma_{x_i \in C_p, x_j \in C_q}$ minimum $\{degOfVar(x_i), degOfVar(x_j)\}$

The best partition among a set of partitions is selected on the basis of goodness of a partition. Goodness measure, *goodness*, of a partition is defined as follows.

Definition 3.11 The goodness of a partition π at level α is defined as follows: $goodness(\pi^{\alpha}) = intra-var(\pi^{\alpha}) + inter-sim(\pi^{\alpha}) - |\pi^{\alpha}|$, where $|\pi^{\alpha}|$ is the number of classes of π.

We have subtracted the term $|\pi^{\alpha}|$ from the sum of intra-class variation and inter-class similarity to remove the bias of goodness value of a partition. Better partition is obtained at higher goodness value. We would like to partition the set of items in Example 3.1 using above goodness measure.

Example 3.4 With reference to Example 3.2, we calculate goodness value of each of the non-trivial partitions.

$intra-var(\pi^{0.001}) = 0.001$, $inter-sim(\pi^{0.001}) = 0.081$, and $|\pi^{0.001}| = 4$. Thus, $goodness(\pi^{0.001}) = -3.916$.
$intra-var(\pi^{0.002}) = 0.003$, $inter-sim(\pi^{0.002}) = 0.06$, and $|\pi^{0.002}| = 3$. Thus, $goodness(\pi^{0.002}) = -2.937$.

The goodness value corresponding to the partition $\pi^{0.002}$ is the largest one. Thus, the partition $\pi^{0.002}$ is the best among the non-trivial partitions. Let us return back to Example 3.1. There are five series of supports corresponding to five items. Based on variation among the supports in a series, we could partition the series as follows: {series 1}, {series 2, series 5}, {series 3, series 4}. Hence, we get the following partition: $\{x_1\}, \{x_2, x_5\}, \{x_3, x_4\}$. The proposed clustering technique also identifies the same partition as the best partition. Thus, it verifies the correctness of the proposed clustering technique. *adv* $(\{x_1\}| \pi^{0.002}) = 0.105$, *adv* $(\{x_2, x_5\}|$ $\pi^{0.002}) = 0.0025$ and *adv* $(\{x_3, x_4\}| \pi^{0.002}) = 0.022$. We find that the average degree of variation of $\{x_2, x_5\}$ is the least among the classes of $\pi^{0.002}$. Thus, the items x_2 and x_5 are most suitable among all the items in the given databases for making strategies of the company. •

We design an algorithm for finding best non-trivial partition of items in multiple databases. First we describe different data structures used in designing an algorithm for determing the best partition of items. For each item there are k supports corresponding to k different years. We maintain $m \times k$ supports for m items in array *supports*. The i-th row of *supports* stores supports corresponding to i-th item for k years, $i = 1, 2,..., m$. Let *means* be a two dimensional array such that the i-th row stores means of supports corresponding to different years for i-th item, $i = 1, 2,..., m$. Let *autocovariances* be a two dimensional array such that the i-th row stores autocovariances of supports corresponding to different lags for i-th item, $i = 1, 2,..., m$. For year j, we compute mean value of supports for year 1 to j. Thus, we get different mean values for different years. Let *stdDevMeans* be the standard deviation of these mean values. For year j, we also compute autocovariances of supports for year 1 to j at different lags. Thus, we get different autocovariances for different lags corresponding to a year. Let *stdDevAutocovars* be the standard deviation of these autocovariances. The degrees of variation of different items are stored in array *degInVar*. Variable S is a one dimensional array containing $^mC_2$ difference in variations. *adv* is a one dimensional array which stores the average degree of variation for the items in each class. The algorithm is presented below (Adhikari et al. 2009).

Algorithm 3.1 Find best non-trivial partition (if it exists) of items in multiple databases.

procedure B*estPartition* (*m*, *supports*)

Inputs:
m: number of items
supports: array of supports of different items corresponding to different years
Outputs:
Best non-trivial partition (if it exists) of items in multiple databases
```
01:  for i = 1 to m do
02:    compute means(i) using formula (3.1) at different years;
03:    let stdDevMeans = standard deviation of mean values for different years;
04:    compute autocovariance(i) using formula (3.2) at different time lags;
05:    let stdDevAutocovars = standard deviation of autocovariances;
06:    compute degOfVar(i) = stdDevMeans + stdDevAutocovar;
07:  end for
08:  for row = 1 to m do
09:    for col = (row + 1) to m do
10:      compute diffInVar(row, col) using formula (3.3);
11:    end for
12:  end for
13:  sort distinct elements in the upper triangle of diffInVar in non-decreasing order into S;
14:  let k = 1; let maxGoodness = -9999; π = φ;
15:  while (k ≤ |S|) do
16:    let curRow = 1; let curClass = 1;
17:    for i = 2 to m do
18:      classLabel(i) = 0;
19:    end for
20:    let classLabel(1) = 1;
21:    let curDiffVar = S(k);
22:    for col = curRow + 1 to m do
23:      if (diffInVar(curRow, col) ≤ curDiffVar) then
24:        if (classLabel (col) = 0) then
25:          classLabel(col) = curClass;
26:        else if (classLabel (col) ≠ curClass) then
27:          partition does not exist at this level;
28:          go to line 49;
29:        end if
30:      end if
31:    end for
```

```
32:    increased curRow by 1;
33:    if (classLabel(curRow) = 0) then
34:      increased curClass by 1;
35:      classLabel(curRow) = curClass;
36:    else curclass = classLabel(curRow);
37:    end if
38:    if (curRow ≤ m) go to line 22; end if
39:    let j = 0;
40:    while ((classLabel(j) ≠ 0) and (j < m)) do
41:      increase j by 1;
42:    end while
43:    if (j = m + 1) then
44:      if (maxGoodness < goodness value of current partition) then
45:        maxGoodness = goodness value of current partition;
46:        store current partition into π;
47:      end if
48:    end if
49:    increase k by1;
50:  end while
51:  return π;
end procedure
```

In this paragraph, we explain the algorithm Algorithm 3.1. It computes *degreeOfVar* for all items using lines 1–7. Matrix *diffInVar* is constructed using lines 8–12. We check the existence of partition at every value in S. We start checking partition by assigning the first item to first class i.e., $curClass = 1$. Also, clustering process is performed row by row, starting from row number 1. At the i-th row, all the items greater than i are classified. During this process, if a labelled item gets another label then we conclude that partition does not exist at the current level. After increasing the current row by 1, we check the class label corresponding to current row. Each row corresponds to an item in the database. If the current row is not labelled yet then the class label is increased by 1. If the goodness value of the current partition is less than the *goodness* value of another partition then current partition is ignored.

Lemma 3.1 *Algorithm 3.1 executes in $O(m^4)$ time.*

Proof Line 2 takes $O(k)$ time to compute *means*(i), for some $i = 1, 2, ..., m$. Also, line 3 takes $O(k)$ time to compute standard deviation of mean values. To compute formula (3.2), we require $O(k)$ time. Thus, line 4 takes $O(k^2)$ time. In line 5, we compute standard deviation of $k - 1$ autocovariance values. Thus, line 5 takes $O(k^2)$ time. The *for-loop* in lines 1–7 repeat m times. Thus, the *for-loop* in lines 1–7 take $O(m \times k^2)$ time. For computing *diffInVar* at a given row and column, it takes $O(1)$ time. Thus, lines 8–12 take $O(m^2)$ time. There are maximum $^{m-1}C_2$ elements in the upper triangle of *diffInVar*. Thus, line 13 takes $O(m^2 \times log(m))$ time. The *while-loop* at line 15 repeats maximum $^{m-1}C_2$ times. Each of the loops at lines 17, 22, and 40 takes $O(m)$ time. To store a partition it takes $O(m)$ time. To compute goodness value for a particular partition, it takes $O(m^2)$ time. Thus, the lines 15–50 take $O(m^4)$ time. The time complexity of *bestPartition* algorithm is $O(m^4)$.

In finding stable items in multiple databases, a class having minimal average degree of variation in the best partition might not be a best class at a given degree of stability. In many applications, we may need to find stable items at a given degree of stability. In this case, it might not be a requirement that the stable items need to form a class of a non trivial partition. Thus, the question of finding a partition might not always arise. To find such a class we follow a different approach.

3.5.2 Finding a Best Class

Before finding a best class, we first define the concept of the best class.

Definition 3.12 Let C be a class of items. C is called a best class at the level of difference in variation α if (1) $|degOfVar(x) - degOfVar(y)| \leq \alpha$, for $x, y \in C$, (2) $adv(C)$ is the minimum among all classes of maximal size, and (3) C has a maximal size.

In Lemma 3.2, we show that it might not be possible to find two classes of maximal size having the same average degree of variation.

Lemma 3.2 *Best class is unique.*

Proof Let $x_1, x_2, \ldots, x_m$ be the items sorted in a non-decreasing degree of variation. Thus the item x_1 has the maximum stability, and the item x_m has the minimum stability. At level α, let the stabilities of items $x_1, x_2, \ldots, x_k$ be less than or equal to α, and the stabilities of items $x_{k+1}, x_{k+2}, \ldots, x_m$ be greater than α, for $1 \leq k \leq m$. The best class has least average degree of variation. Also, the difference in variation of two items in the class is less than or equal to α. Thus, $\{x_1, x_2, \ldots, x_k\}$ forms the best class. We are not concerned whether it becomes a member of a partition. $adv(\{x_1, x_2, \ldots, x_k\})$ is the minimum, and hence the best class is unique.

We might be interested in finding best class of items in multiple databases. We use array *class* to hold the best class of items. In the following, we provide an algorithm in finding best class of items in multiple databases.

Algorithm 3.2. Find the best class of items in multiple databases induced by stability.

Inputs:
m: number of items
α: level of degree of variation
supports: array of supports of different items corresponding to different years
Outputs:
Best class of items in multiple databases
01: perform lines 01 – 07 of Algorithm 3.1;
02: sort array *degOfVar* in non-decreasing order;
03: **let** $class(1) = degOfVar(1)$; **let** $count = 1$; **let** $avgVar = 0$;
04: **for** $i = 2$ to m **do**
05:    $class(i) = -1$;
06: **end for**
07: **for** i = 2 to m **do**
08:    **if** $((degOfVar(i) - degOfVar(1)) \leq \alpha)$
09:       $class(i) = degOfVar(i)$;
10:       increase *count* by 1;
11:       $avgVar = avgVar + degOfVar(i)$;
12: **end for**
13: $avgVar = avgVar / count$;
14: return (*class, count, avgVar*);
end procedure

procedure B*estClass* (*m*, α, *supports*)

Let us elaborate on the individual lines of the above algorithm. We compute degree of variations for all items using lines 1–7 and store them in array *degreeOfVar* in a non-decreasing order. The best class would contain the first item of *degreeOfVar*. The item with least *degreeOfVar* is assigned to *class* 1. An item *i* is included in the best class if $(degOfVar(i) - degOfVar(1)) \leq α$. Algorithm 3.2 returns best class *class*, the number of items in the best class *count*, and the average degree of variation (*avgVar*) of the best class.

Lemma 3.3 *Algorithm 3.2 executes in maximum* $\{O(m \times k^2), O(m \times \log(m))\}$ *time.*

Proof Line 1 executes in $O(m \times k^2)$ time [Lemma 3.2], where *k* is the number of years. There are two for loops in Algorithm 3.2 apart from loops placed in line 1. Each of these loops executes in $O(m)$ time. Line 2 takes $O(m \times \log(m))$ time. Thus, the lemma follows.

3.6 Experiments

We have carried out a number of experiments to study the effectiveness of our approach. All the experiments have been implemented on a 1.6 GHz Pentium IV with 256 MB of memory, using visual C++ (version 6.0). We present the experimental results using two real datasets *mushroom* (Frequent itemset mining dataset repository 2004), and *ecoli* (UCI ML repository). Dataset *ecoli* is a subset of *ecoli database* and it has been processed for the purpose of conducting experiments. *Random*-68 is a synthetic database, which has been generated for the purpose of conducting experiments.

Let *DB*, *NT*, *ALT*, *AFI*, and *NI* denote database, the number of transactions, average length of a transaction, average frequency of an item, and number of items respectively. We present some characteristics of these datasets in Table 3.1. Each dataset has been divided into ten databases, called input databases, for the purpose of conducting experiments. The input databases obtained from *mushroom* and *ecoli* are named as M_i, and E_i, $i = 0, 1, ..., 9$. We present some characteristics of the input databases in Table 3.2. In Table 3.3, we present top 10 stable items encountered in multiple databases.

Table 3.1 Dataset characteristics

Dataset	*NT*	*ALT*	*AFI*	*NI*
Mushroom	8124	24.00	1624.80	120
Ecoli	336	7.00	25.84	91
Random-68	3000	5.46	280.99	68

Table 3.2 Time database characteristics

DB	NT	ALT	AFI	NI	DB	NT	ALT	AFI	NI
M_0	812	24.00	295.27	66	$M5$	812	24.00	221.45	88
M_1	812	24.00	286.59	68	$M6$	812	24.00	216.53	90
M_2	812	24.00	249.85	78	$M7$	812	24.00	191.06	102
M_3	812	24.00	282.43	69	$M8$	812	24.00	229.27	85
M_4	812	24.00	259.84	75	$M9$	816	24.00	227.72	86
E_0	33	7.00	4.62	50	$E5$	33	7.00	3.92	59
E_1	33	7.00	5.13	45	$E6$	33	7.00	3.50	66
E_2	33	7.00	5.50	42	$E7$	33	7.00	3.92	59
E_3	33	7.00	4.81	48	$E8$	33	7.00	3.40	68
E_4	33	7.00	3.40	68	$E9$	39	7.00	4.55	60
R_0	300	5.59	28.68	68	$R5$	300	5.14	26.68	68
R_1	300	5.42	28.00	68	$R6$	300	5.51	28.35	68
R_2	300	5.36	27.65	68	$R7$	300	5.50	28.34	68
R_3	300	5.54	28.46	68	$R8$	300	5.54	28.47	68
R_4	300	5.53	28.38	68	$R9$	300	5.48	28.24	68

Table 3.3 Top 10 stable items in multiple databases

Mushroom		Ecoli		Random-68	
Item	DegOfVar	Item	DegOfVar	Item	DegOfVar
85	0.0000	1	0.0013	42	0.0012
8	0.0002	99	0.0018	41	0.0018
12	0.0003	91	0.0023	37	0.0023
75	0.0003	4	0.0025	67	0.0027
89	0.0003	94	0.0036	11	0.0028
62	0.0009	15	0.0039	45	0.0029
22	0.0010	12	0.0039	18	0.0030
20	0.0011	19	0.0040	56	0.0031
82	0.0012	3	0.0040	3	0.0031
33	0.0014	10	0.0044	28	0.0031

The best partition of items in *mushroom* dataset is obtained at level 0.25. The intra variation, inter similarity, and goodness value are 413.60, 377.58, and 789.18, respectively. The best partition contains two classes. The best class is given as follows: {1, 2, 3, 4, 5, 6, 7, 8, 9, 10, 11, 12, 13, 14, 15, 16, 17, 18, 19, 20, 21, 22, 23, 24, 25, 26, 27, 28, 29, 30, 31, 32, 33, 34, 35, 36, 37, 38, 39, 40, 41, 42, 43, 44, 45, 46, 47, 48, 49, 50, 51, 52, 53, 54, 55, 57, 58, 59, 60, 61, 62, 63, 64, 65, 66, 67, 68, 69, 70, 71, 72, 73, 74, 75, 76, 77, 78, 79, 80, 81, 82, 83, 84, 85, 86, 87, 88, 89, 90, 91, 92, 93, 94, 95, 96, 97, 98, 99, 100, 101, 102, 103, 104, 105, 106, 107, 108, 109, 110, 111, 112, 113, 114, 115, 116, 117, 118, 119}. It has average degree of variation 0.06.

The best partition of items in *ecoli* dataset is obtained at level 0.06. The amount of intra variation, inter similarity, and goodness value are 44.18, 185.61, and

227.79 respectively. The best partition contains two classes. The best class is given as follows: {0, 1, 3, 4, 5, 6, 7, 8, 10, 11, 12, 14, 15, 16, 17, 18, 19, 20, 21, 22, 23, 24, 25, 26, 27, 28, 29, 30, 31, 32, 33, 34, 36, 37, 38, 39, 40, 41, 43, 45, 46, 47, 48, 49, 50, 51, 52, 53, 54, 55, 56, 57, 58, 59, 60, 61, 62, 63, 64, 65, 66, 67, 69, 70, 71, 72, 73, 74, 75, 76, 77, 78, 79, 80, 81, 82, 83, 84, 85, 86, 87, 88, 89, 90, 91, 92, 94, 99, 100}. It has average degree of variation 0.02.

The best partition of items in *random*-68 dataset is obtained at the level 0.02. The amount of intra variation, inter similarity, and goodness value are 4.63, 18.61, and 21.24, respectively. The best partition contains two classes. The best class is given as follows: {1, 2, 3, 5, 6, 7, 8, 9, 10, 11, 12, 13, 14, 15, 16, 17, 18, 19, 20, 21, 22, 23, 24, 25, 26, 27, 28, 29, 30, 31, 32, 33, 34, 35, 36, 37, 38, 39, 40, 41, 42, 43, 44, 45, 46, 47, 48, 49, 50, 51, 52, 53, 54, 55, 56, 57, 58, 59, 60, 61, 62, 63, 64, 65, 66, 67, 68}. It has average degree of variation 0.01.

In Tables 3.4, 3.5 and 3.6 we present five best classes and their average degree of variations for a given value of α for each database.

Table 3.4 Five best classes in multiple databases (*mushroom*)

α	Items	adv
0.0002	{85, 8}	0.00009
0.0003	{85, 8, 12, 75, 89}	0.00021
0.0010	{85, 8, 12, 75, 89, 62}	0.00033
0.0012	{85, 8, 12, 75, 89, 62, 22, 20}	0.00051
0.0014	{85, 8, 12, 75, 89, 62, 22, 20, 82}	0.00059

Table 3.5 Five best classes in multiple databases (*ecoli*)

α	Items	adv
0.0020	{1, 99, 91, 4}	0.00196
0.0025	{1, 99, 91, 4, 94}	0.00230
0.0030	{1, 99, 91, 4, 94, 15, 12, 19, 3}	0.00303
0.0035	{1, 99, 91, 4, 94, 15, 12, 19, 3, 10, 6}	0.00331
0.0050	{1, 99, 91, 4, 94, 15, 12, 19, 3, 10, 6, 18}	0.00354

Table 3.6 Five best classes in multiple databases (*random*-68)

α	Items	adv
0.0010	{42, 41}	0.00152
0.0015	{42, 41, 37}	0.00178
0.0017	{42, 41, 37, 67, 11, 45}	0.00229
0.0020	{42, 41, 37, 67, 11, 45, 18, 56, 3, 28}	0.00261
0.0022	{42, 41, 37, 67, 11, 45, 18, 56, 3, 28, 7, 53}	0.00271

Fig. 3.2 Execution time versus number of data sources for *mushroom*

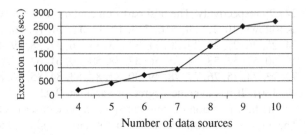

Fig. 3.3 Execution time versus number of data sources for *ecoli*

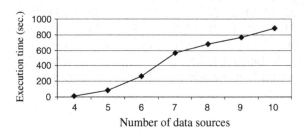

Fig. 3.4 Execution time versus number of data sources for *random*-68 data

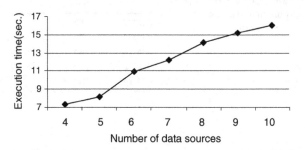

We have analyzed execution time with respect to number of data sources. We observe in Figs. 3.2, 3.3 and 3.4 that the execution time increases as the number of data sources increases.

3.7 Conclusions

Many organizations deal with a large number of items collected for a long period of time. In this scenario, data analyses are carried out for the data items dealt with. Studying stability of an item is an important issue, since there are a series of time databases. For the purpose of conducting experiments, we have considered yearly databases. Thus, the notion of stability of an item is introduced here. Stable items are useful for modelling various strategies of an organization. We design an algorithm for clustering items in multiple databases based on degree of stability. Afterwards we have introduced the notion of best class in a cluster. We have designed an algorithm to find the best class in a cluster. The best class is obtained by considering an average degree of variation in a class. A class having the least average degree variation is the best one. We have conducted experiments on three datasets, and provided detailed data analyses related to stability of an item.

References

Adhikari A (2013) Clustering local frequency items in multiple databases. Inf Sci 237:221–241

Adhikari A, Rao PR (2008a) Synthesizing heavy association rules from different real data sources. Pattern Recogn Lett 29(1):59–71

Adhikari A, Rao PR (2008b) Efficient clustering of databases induced by local patterns. Decis Support Syst 44(4):925–943

Adhikari J, Rao PR, Adhikari A (2009) Clustering items in different data sources induced by stability. Int Arab J Inf Technol 6(4):394–402

Adhikari A, Ramachandrarao P, Pedrycz W (2010) Developing multi-databases mining applications. Springer, London

Agrawal R, and Srikant R (1994) Fast algorithms for mining association rules. In: Proceedings of 20th very large databases (VLDB) conference, pp 487–499

Agrawal R, Imielinski T, Swami A (1993) Mining association rules between sets of items in large databases. In: Proceedings of ACM SIGMOD conference management of data, pp 207–216

Albanese M, Molinaro C, Persia F, Picariello A, Subrahmanian VS (2013) Discovering the top-k "unexplained" sequences in time-stamped observation data. IEEE Trans Knowl Data Eng 9(1):1

Bluman AG (2006) Elementary statistics: a step by step approach. Mcgraw Hill, New York

Brockwell J, Richard D (2002) Introduction to time series and forecasting. Springer, Berlin

Estivill-Castro V, Yang J (2004) Fast and robust general purpose clustering algorithms. Data Min Knowl Disc 8(2):127–150

Frequent itemset mining dataset repository (2004) http://fimi.cs.helsinki.fi/data

Guil F, Marín R (2012) A tree structure for event-based sequence mining. Knowl-Based Syst 35:186–200

Han J, Pei J, Yiwen Y (2000) Mining frequent patterns without candidate generation. In: Proceedings of ACM SIGMOD conference management of data, pp 1–12

Jain AK, Murty MN, Flynn PJ (1999) Data clustering: areview. ACM Comput Surv 31(3):264–323

Leonard M, Wolfe B (2005) Mining transactional and time series data. In: Proceedings of SUGI 30, pp 080–30

Liu CL (1985) Elements of discrete mathematics. McGraw-Hill, New York

Liu B, Ma Y, Lee R (2001) Analyzing the interestingness of association rules from the temporal dimension. In: Proceedings of IEEE international conference on data mining, pp 377–384

Matsubara Y, Sakurai Y, Faloutsos C, Iwata T, Yoshikawa M (2012) Fast mining and forecasting of complex time-stamped events. In: Proceedings of KDD conference, pp 271–279

Tan PN, Kumar V, Srivastava J (2002) Selecting the right interestingness measure for association patterns. In: Proceedings of SIGKDD conference, pp 32–41

UCI ML repository content summary. http://www.ics.uci.edu/~mlearn/MLSummary.html

Wu X, Zhang C, Zhang S (2005) Database classification for multi-database mining. Inf Syst 30(1):71–88

Yang K, Shahabi C (2005) On the stationarity of multivariate time series for correlation-based data. In: Proceedings of ICDM, pp 805–808

Zhang T, Ramakrishnan R, Livny M (1997) BIRCH: a new data clustering algorithm and its applications. Data Min Knowl Disc 1(2):141–182

Zhang S, Wu X, Zhang C (2003) Multi-database mining. IEEE Comput Intell Bull 2(1):5–13

Zhang S, Zhang C, Wu X (2004) Knowledge discovery in multiple databases. Springer, London

Chapter 4
Synthesizing Global Patterns in Multiple Large Data Sources

Effective data analysis using multiple databases requires patterns that are almost error-free. As the local pattern analysis might extract patterns of low quality from multiple databases, it becomes necessary to improve mining multiple databases. In this chapter, we present an idea of multi-database mining by making use of local pattern analysis. We elaborate on the existing specialized and generalized techniques which are used for mining multiple large databases. In the sequel, we discuss a certain generalized technique, referred to as a pipelined feedback model, which is of particular relevance for mining multiple large databases. It significantly improves the quality of the synthesized global patterns. We define two types of error occurring in multi-database mining techniques. Experimental results are reported for both real-world and synthetic databases. They help us assess the effectiveness of the pipelined feedback model.

4.1 Introduction

As noted earlier, many large companies operate from a number of branches usually located at different geographical regions. Each branch collects data continuously and local data become stored locally. The collection of all branch databases might be large. Many corporate decisions of a multi-branch company are based on data stored over the branches. The challenges are to make meaningful decisions, which are based on large volume of distributed data. This creates not only risk but also offers opportunities. One of the risks is a significant amount investment on hardware and software to deal with multiple large databases. The use of inefficient data mining techniques has to be taken into account and in many scenarios this shortcoming could be very detrimental to the quality of results.

Based on the number of data sources, patterns in multiple databases could be classified into three categories. These are local patterns, global patterns and patterns that are neither local nor global. A pattern based on a single database is called a local pattern. Local patterns are useful for local data analysis, and locally restricted decision making activities (Adhikari and Rao 2008b; Wu et al. 2005;

A. Adhikari et al., *Data Analysis and Pattern Recognition in Multiple Databases*,
Intelligent Systems Reference Library 61, DOI: 10.1007/978-3-319-03410-2_4,
© Springer International Publishing Switzerland 2014

Zhang et al. 2004b). On the other hand, global patterns are based on all the databases taken into consideration. They are useful for data analyses of global nature (Adhikari and Rao 2008a; Wu and Zhang 2003; Adhikari 2012, 2013; Zhang et al. 2009) and global decision making problems. The intent of this chapter is to introduce and analyze a certain global model of data mining, referred to as a pipelined feedback technique (PFT) (Adhikari et al. 2010) which is used for mining/synthesizing global patterns in multiple large databases.

In Sect. 4.2, we formalize the idea of multi-database mining using local pattern analysis. Next, we discuss existing generalized multi-database mining techniques (Sect. 4.3). We analyze the existing specialized multi-database mining techniques in Sect. 4.4. The pipelined feedback technique for mining multiple large databases is covered in Sect. 4.5. We also define a way in which an error associated with the model is quantified (Sect. 4.6). In Sect. 4.7, we provide experimental results using both synthetic and real-world databases.

4.2 Multi-Database Mining Using Local Pattern Analysis

Consider a large company that deals with multiple large databases. For mining multiple databases, we are faced with three scenarios viz., (1) Each of the local databases is small, so that a single database mining technique (SDMT) could mine the union of all databases. (2) At least one of the local databases is large, so that a SDMT could mine every local database, but fail to mine the union of all local databases. (3) At least one of the local databases is very large, so that a SDMT fails to mine at least one local database. We are faced with challenges when handling cases (2) and (3) and these challenges are inherently present because of the large size of some databases.

The first question, which comes to our mind is whether a traditional data mining technique (Agrawal and Srikant 1994; Han et al. 2000; Coenen et al. 2004) could provide a sound solution when dealing with multiple large databases. To apply a "traditional" data mining technique we need to amass all the branch databases together. In such cases, a traditional data mining approach might not offer a sound solution due to the following reasons:

- The approach might not be suitable as it requires heavy investment on hardware and software to deal with a large volume of data.
- A single computer might take unreasonable amount of time to mine huge masses of data.
- It is difficult to identify local patterns if a "traditional" data mining technique is applied to the collection of all local databases.

In light of these problems and associated constraints, as encountered so far there have been attempts to deal with multi-database mining in a different way.

Zhang et al. (2003) designed a multi-database mining technique (MDMT) using local pattern analysis. Multi-database mining using local pattern analysis could be classified into two categories viz., the techniques that analyze local patterns, and the techniques that analyze approximate local patterns. A multi-database mining technique using local pattern analysis could be viewed as a two-step process, denoted symbolically as M+S. Its essence can be explained as follows:

- Mine each local database using a SDMT by applying the model M (Step 1)
- Synthesize patterns using the algorithm S (Step 2).

We use the notation of M+S to stress a character of a multi-database mining technique in which we first use the model of mining (M) being followed by the synthesizing algorithm S.

One could apply sampling techniques (Babcock et al. 2003) for taming large volume of data. If an itemset is frequent in a large dataset then it is likely to be frequent in a sample dataset. Thus, one can mine patterns approximately in a large dataset by analyzing patterns in a representative sample dataset.

In addition to generalized multi-database mining techniques, there exist also specialized multi-database mining techniques. In what follows, we discuss some of the existing multi-database mining techniques.

4.3   Generalized Multi-Database Mining Techniques

There is a significant variety of techniques that can be used in multi-database mining applications.

4.3.1   Local Pattern Analysis

Under this model of mining multiple databases, each branch requires to mine the database using a traditional data mining technique. Afterwards, each branch is required to forward the pattern base to the central office. Then the central office processes the locally processed pattern bases collected from different branches to synthesize the global patterns and subsequently to support decision-making activities. Zhang et al. (2003) designed a multi-database mining technique (MDMT) using local pattern analysis. In Chap. 1, we presented this model in detail. We have proposed an extended model of local pattern analysis (Adhikari and Rao 2008a). It improves the quality of synthesized global patterns in multiple databases. In addition, it supports a systematic approach to synthesize the global patterns. In Chap. 2, we have presented the extended model of local pattern analysis for mining multiple large databases.

4.3.2 Partition Algorithm

For the purpose of mining multiple databases, one could apply *partition algorithm* (PA) proposed by Savasere et al. (1995). In Chap. 1, we have discussed this algorithm.

4.3.3 IdentifyExPattern Algorithm

Zhang et al. (2004a) have proposed algorithm, *IdentifyExPattern* (IEP) for identifying global exceptional patterns in multi-databases. Every local database is mined separately at *random order* (RO) using a SDMT to synthesize global exceptional patterns. For identifying global exceptional patterns in multiple databases, the following pattern synthesizing approach has been proposed. A pattern in a local database is assumed as absent, if it does not get reported. Let $supp_a(p, DB)$ and $supp_s(p, DB)$ be the actual (i.e., apriori) support and synthesized support of pattern p in database DB. Let D be the union of all local databases. Then support of pattern p has been synthesized in D based on the following expression:

$$supp_s(p, D) = \frac{1}{num(p)} \sum_{i=1}^{num(p)} (supp_a(p, D_i) - \alpha)/(1 - \alpha) \qquad (4.1)$$

where $num(p)$ is the number of databases that report p at a given minimum support level (α).

The size (i.e., the number of transactions) of a local database and support of an itemset in a local database seem to be important parameters that are used to determine the presence of an itemset in a database, since the number of transactions containing the itemset X in a database D_1 is equal to $supp(X, D_1) \times size(D_1)$. The major concern in this investigation is that the algorithm does not consider the size of a local database to synthesize the global support of a pattern. Using the IEP algorithm, the global support of a pattern has been synthesized using only supports of the pattern present in local databases.

4.3.4 RuleSynthesizing Algorithm

Wu and Zhang (2003) have proposed *RuleSynthesizing* (RS) algorithm for synthesizing high-frequent association rules in multiple databases. Using this technique, every local database is mined separately at *random order* (RO) using a SDMT for synthesizing high-frequency association rules. A pattern in a local database is assumed as absent, if it does not get reported. Based on the association rules present in different databases, the authors have estimated weights of different databases. Let w_i be the weight of the i-th database, $i = 1, 2, ..., n$. Without any loss of generality, let the association rule r be extracted from first m databases, for $1 \leq m \leq n$. Here, $supp_a(r, D_i)$ has been assumed to be equal to 0, for $i = m + 1$, $m + 2, ..., n$. Then the support of r in D has been determined in the following way:

$$supp_s(r, D) = w_1 \times supp_a(r, D_1) + \cdots + w_m \times supp_a(r, D_m) \qquad (4.2)$$

Algorithm *RuleSynthesizing* offers an indirect approach to synthesizing association rules in multiple databases. Thus the time complexity of the algorithm is reasonably high. The algorithm executes in $O(n^4 \times maxNosRules \times totalRules^2)$ time, where n, *maxNosRules*, and *totalRules* are the number of data sources, the maximum among the numbers of association rules extracted from different databases, and the total number of association rules in different databases, respectively.

4.4 Specialized Multi-Database Mining Techniques

For finding solution to a specific application, it might be possible to devise a better multi-database mining technique. In this section, we elaborate in detail on three specific multi-database mining techniques.

4.4.1 Mining Multiple Real Databases

We have proposed algorithm *Association-Rule-Synthesis* (ARS) for synthesizing association rules in multiple real data sources (Adhikari and Rao 2008a). The algorithm uses the model shown in Fig. 2.1. While synthesizing an association rule, it uses a specific method whose functioning can be explained as follows: For real databases, the trend of the customers behaviour exhibited in a single database is usually present in other databases. In particular, a frequent itemset in one database is usually present in some transactions of other databases even if it does not get extracted. The proposed estimation procedure captures such trend and estimates the support of a missing association rule. Without any loss of generality, let the itemset X be extracted from first m databases, $1 \leq m \leq n$. Then trend of X in first m databases could be expressed as follows:

$$trend^{1, m}(X|\alpha) = \frac{1}{\sum_{i=1}^{m} |D_i|} \times \sum_{i=1}^{m} (supp_a(X, D_i) \times |D_i|) \qquad (4.3)$$

The number of transactions in a database could be considered as its weight. In (4.3), the trend of X in first m databases is estimated as a weighted sum of supports in the first m databases. We can use the detected trend of X encountered in the first m databases for synthesizing support of X in D. We estimate the support of X in database D_j by computing the expression $\alpha \times trend^{1,\,m}(X \mid \alpha)$, $j = k + 1, k + 2,$ ..., n. Then the synthesized support of X is determined as follows:

$$\text{supp}_s(X, D) = \frac{trend^{1,\,m}(X \mid \alpha)}{\sum_{i=1}^{n} |D_i|} \times \left[(1 - \alpha) \times \sum_{i=1}^{m} |D_i| + \alpha \times \sum_{i=1}^{n} |D_i| \right] \quad (4.4)$$

Association-Rule-Synthesis algorithm might return approximate global patterns, since an itemset might not get extracted from all the databases.

4.4.2 Mining Multiple Databases for the Purpose of Studying a Set of Items

Many important decisions are based on a set of specific items called the *select items*. We have proposed a technique for mining patterns of select items in multiple databases (Adhikari et al. 2011). The details of the technique are presented in Chap. 6.

4.4.3 Study of Temporal Patterns in Multiple Databases

Adhikari et al. (2009) have proposed a technique for clustering items in multiple databases based on their level of stability where a certain stability measure is used to quantify this feature. The technique has been presented in Chap. 3.

4.5 Mining Multiple Databases Using Pipelined Feedback Technique

Before applying the pipelined feedback technique, one needs to prepare data warehouses at different branches of a multi-branch organization. In Fig. 2.1, we have shown how to preprocess data warehouse at each branch. Let W_i be the data warehouse corresponding to the i-th branch, $i = 1, 2, ..., n$. Then the local patterns for the i-th branch are extracted from W_i, $i = 1, 2, ..., n$. We mine each data warehouse using any SDMT. In Fig. 4.1, we present a model of mining multiple databases (Adhikari et al. 2010).

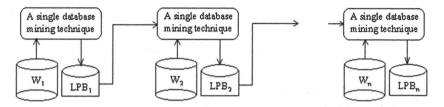

Fig. 4.1 Pipelined feedback technique of mining multiple databases

In PFT, W_1 is mined using a SDMT and as result a local pattern base LPB_1 becomes extracted. While mining W_2, all the patterns in LPB_1 are extracted irrespective of their values of interestingness measures like, minimum support and minimum confidence. Apart from these patterns, some new patterns that satisfy user-defined threshold values of interestingness measures are also extracted. In general, while mining W_i, all the patterns in W_{i-1} are mined irrespective of their values of interestingness measures, and some new patterns that satisfy user-defined threshold values of interestingness measures, $i = 2, 3, ..., n$. Due to this nature of mining each data warehouse, the technique is called a feedback model. Thus, $|LPB_{i-1}| \leq |LPB_i|$, for $i = 2, 3, ..., n$. There are $n!$ arrangements of pipelining for n databases. All the arrangements of data warehouses might not produce the same result of mining. If the number of local patterns increases, one gets more accurate global patterns which leads to a better analysis of local patterns. An arrangement of data warehouses would produce near optimal result if the cardinality $|LPB_n|$ is maximal. Let $size(W_i)$ be the size of W_i (in bytes), $i = 1, 2, ..., n$. We adhere to the following rule of thumb regarding the arrangements of data warehouses for the purpose of mining. The number of patterns in W_i is greater than or equal to the number of patterns in W_{i-1} when $size(W_i) \geq size(W_{i-1})$, $i = 2, 3, ..., n$. For the purpose of increasing the number of local patterns, W_{i-1} precedes W_i in the pipelined arrangement of mining data warehouses if $size(W_{i-1}) \geq size(W_i)$, $i = 2, 3, ..., n$. Finally, we analyze the patterns in $LPB_1, LPB_2, ...,$ and LPB_n to synthesize global patterns, or analyze local patterns.

Let W be the collection of all branch data warehouses. For synthesizing global patterns in W we discuss here a simple pattern synthesizing (SPS) algorithm. Without any loss of generality, let the itemset X be extracted from first m databases, for $1 \leq m \leq n$. Then the synthesized support of X in W comes in the following form

$$supp_s(X, W) = \frac{1}{\sum_{i=1}^{n} |W_i|} \times \sum_{i=1}^{m} [supp_a(X, W_i) \times |W_i|] \qquad (4.5)$$

4.5.1 Algorithm Design

In this section, we present an algorithm for mining multiple large databases. The method is based on the pipelined feedback model discussed above.

Algorithm 4.1 Mine multiple data warehouses using pipelined feedback model.
procedure *PipelinedFeedback Technique* $(W_1, W_2, ..., W_n)$
Input: $W_1, W_2, ..., W_n$
Output: local pattern bases
01: **for** $i = 1$ to n **do**
02:    **if** W_i does not fit in memory **then**
03:       partition W_i into $W_{i1}, W_{i2}, ...,$ and Wip_i for an integer p_i;
04:    **else** $W_{i1} = W_i$;
05:    **end if**
06: **end for**
07: sort data warehouses on size in non-increasing order and the data warehouses
     are renamed as $DW_1, DW_2, ..., DW_N$, where $N = \sum_{i=1}^{n} p_i$;
08: **let** $LPB_0 = \phi$,
09: **for** $i = 1$ to N **do**
10:    mine DW_i using a SDMT with input LPB_{i-1};
11: **end for**
12: **return** $LPB_1, LPB_2, ..., LPB_N$;
end procedure

In the algorithm, the usage of LPB_{i-1} during mining DW_i has been explained above. Once a pattern has been extracted from a data warehouse, then it also gets extracted from the remaining data warehouses. Thus, the algorithm *Pipelined-Feedback Technique* improves the quality of synthesized patterns as well as contributes significantly to an analysis of local patterns.

4.6 Error Evaluation

To evaluate the quality of MDMT: PFT+SPS, one needs to quantify the error produced by the method. First, in an experiment we mine frequent itemsets in multiple databases using PFT, and afterwards synthesize global patterns using the SPS algorithm. We have to find how the global synthesized support differs from the exact (apriori) support of an itemset.

PFT improves mining multiple databases significantly over local pattern analysis. In the PFT, we have $LPB_{i-1} \subseteq LPB_i$, for $i = 2, 3, ..., n$. Then, patterns in $LPB_i - LPB_{i-1}$ are generated from databases $D_i, D_{i+1}, ..., D_n$. We assume $supp_a(X, D_j) = 0$, for each $X \in LPB_i - LPB_{i-1}$, and $j = 1, 2, ..., i - 1$. Thus, the error of mining X could be defined as follows:

$$E(X|\text{PFT, SPS}) = \left| supp_a(X,\ D) - \frac{1}{\sum_{j=1}^{n} |D_j|} \times \sum_{j=i}^{n} \left[supp_a(X, D_j) \times |D_j| \right] \right|,$$

$$\text{for } X \in LPB_i - LPB_{i-1} \quad \text{and} \quad i = 2, 3, \ldots, n.$$

Also,

$$E(X|\text{PFT, SPS}) = 0, \text{ for } X \in LPB_1. \tag{4.6}$$

When a frequent itemset is reported from D_1 then it gets reported from every databases using PFT. Thus, $E(X|\text{PFT, SPS}) = 0$, for $X \in LPB_1$.

Otherwise, an itemset X is not reported from all the databases. It is synthesized using SPS algorithm. Then the synthesized support is subtracted from its apriori support for finding the error of mining X.

There are several ways one could define the error of an experiment. In particular, one could concentrate on the following definitions.

1. *Average error* (AE)

$$AE(D, \alpha) = \frac{1}{|LPB_1 + \sum_{i=2}^{n} (LPB_i - LPB_{i-1})|} \sum_{X \in [LPB_1 + \cup \left\{ \bigcup_{i=2}^{n} (LPB_i - LPB_{i-1}) \right\}]} E(X|\text{PFT, SPS})]$$

$$\tag{4.7}$$

2. *Maximum error* (ME)

$$ME(D, \alpha) = maximum \left\{ E(X|\text{PFM, SPS}), \text{ for } X \in LPB_1 \bigcup \left\{ \bigcup_{i=2}^{n} (LPB_i - LBP_{i-1}) \right\} \right\} \tag{4.8}$$

where $supp_a(X_i, D)$ is obtained by mining D using a traditional data mining technique, $i = 1, 2, \ldots, m$. $supp_s(X_i, D)$ is obtained by SPS, for $i = 1, 2, \ldots, m$.

4.7 Experiments

We have carried out a series of experiments to study and quantify the effectiveness of the PFM. We present experimental results using three synthetic databases and two real-world databases. The synthetic databases are *T10I4D100K* (*T*) (Frequent itemset mining dataset repository 2004), *random500* (*R1*) and *random1000* (*R2*). The databases *random500* and *random1000* are generated synthetically for the purpose of conducting experiments. The real databases are *retail* (*R*) (Frequent itemset mining dataset repository 2004) and *BMS-Web-Wiew-1* (*B*) (Frequent itemset mining dataset repository 2004). The main characteristics of these datasets are displayed in Table 4.1.

Table 4.1 Database characteristics

D	NT	ALT	AFI	NI
T	100,000	11.10	1276.12	870
R	88,162	11.31	99.67	10,000
B	149,639	2.00	155.71	1,922
R1	10,000	6.47	109.40	500
R2	10,000	12.49	111.86	1,000

Let NT, AFI, ALT, and NI denote the number of transactions, average frequency of an item, average length of a transaction, and number of items in a database, respectively. Each of the above databases is split into 10 databases for the purpose of carrying out experiments. The databases obtained from T, R, B, $R1$ and $R2$ are named as T_i, R_i, B_i, $R1_i$ and $R2_i$, respectively, for $i = 0, 1, ..., 9$. The databases T_i, R_i, B_i, $R1_i$, $R2_i$ are called input databases (DBs), $i = 0, 1, ..., 9$. Some characteristics of these input databases are presented in the Table 4.2. In Tables 4.3 and 4.4, we include some outcomes to quantify how the proposed technique improves the results of mining. We have completed experiments using other MDMTs on these databases for the purpose of comparing them with MDMT: PFT+SPS.

Table 4.2 Input database characteristics

D	NT	ALT	AFI	NI	DB	NT	ALT	AFI	NI
T_0	10,000	11.06	127.66	866	T_5	10,000	11.14	128.63	866
T_1	10,000	11.133	128.41	867	T_6	10,000	11.11	128.56	864
T_2	10,000	11.07	127.64	867	T_7	10,000	11.10	128.45	864
T_3	10,000	11.12	128.44	866	T_8	10,000	11.08	128.56	862
T_4	10,000	11.14	128.75	865	T_9	10,000	11.08	128.11	865
R_0	9,000	11.24	12.07	8,384	R_5	9,000	10.86	16.71	5,847
R_1	9,000	11.21	12.27	8,225	R_6	9,000	11.20	17.42	5,788
R_2	9,000	11.34	14.60	6,990	R_7	9,000	11.16	17.35	5,788
R_3	9,000	11.49	16.66	6,206	R_8	9,000	12.00	18.69	5,777
R_4	9,000	10.96	16.04	6,148	R_9	7,162	11.69	15.35	5,456
B_0	14,000	2.00	14.94	1,874	B_5	14,000	2.00	280.00	100
B_1	14,000	2.00	280.00	100	B_6	14,000	2.00	280.00	100
B_2	14,000	2.00	280.00	100	B_7	14,000	2.00	280.00	100
B_3	14,000	2.00	280.00	100	B_8	14,000	2.00	280.00	100
B_4	14,000	2.00	280.00	100	B_9	23,639	2.00	472.78	100
$R1_0$	1,000	6.37	10.73	500	$R1_5$	1,000	6.34	10.68	500
$R1_1$	1,000	6.50	11.00	500	$R1_6$	1,000	6.62	11.25	500
$R1_2$	1,000	6.40	10.80	500	$R1_7$	1,000	6.42	10.83	500
$R1_3$	1,000	6.52	11.05	500	$R1_8$	1,000	6.58	11.16	500
$R1_4$	1,000	6.30	10.60	500	$R1_9$	1,000	6.65	11.30	500
$R2_0$	1,000	6.42	5.43	996	$R2_5$	1,000	6.44	5.46	997
$R2_1$	1,000	6.41	5.44	995	$R2_6$	1,000	6.48	5.50	996
$R2_2$	1,000	6.56	5.58	995	$R2_7$	1,000	6.48	5.49	997
$R2_3$	1,000	6.53	5.54	998	$R2_8$	1,000	6.54	5.56	996
$R2_4$	1,000	6.50	5.54	991	$R2_9$	1,000	6.50	5.56	988

Table 4.3 Error obtained for the first three databases for selected value of α

Database	T10I4D100K		Retail		BMS-Web-Wiew-1	
α	0.05		0.11		0.19	
Error type	AE	ME	AE	ME	AE	ME
MDMT: RO+IEP	0.01	0.04	0.01	0.06	0.05	0.15
MDMT: RO+RS	0.01	0.04	0.01	0.06	0.02	0.13
MDMT: RO+ARS	0.01	0.04	0.01	0.06	0.02	0.11
MDMT: PFT+SPS	0	0.05	0.01	0.06	0	0
MDMT: RO+PA	0	0	0	0	0	0

Table 4.4 Error reported for the last two databases for selected value of α

Database	Random500		Random1000	
α	0.005		0.004	
Error type	AE	ME	AE	ME
MDMT: RO+IEP	0.01	0.01	0.01	0.01
MDMT: RO+RS	0.01	0.01	0	0.01
MDMT: RO+ARS	0.01	0.01	0.01	0.01
MDMT: PFT+SPS	0.01	0.01	0	0
MDMT: RO+PA	0	0	0	0

Fig. 4.2 AE versus α for experiments conducted for database T

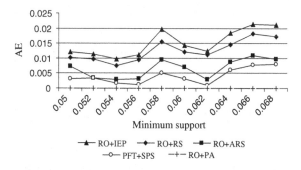

Fig. 4.3 AE versus α for experiments using database R

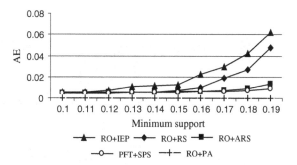

Figures 4.2, 4.3, 4.4, 4.5 and 4.6 show values of average error versus different values of α. From these graphs, we conclude that AE typically increases as α increases. The number of databases reporting a pattern decreases when the values of α increase. Thus, the AE of synthesizing patterns normally increases as α increases. In case of Fig. 4.4, the graphs for MDMT: PFT+SPS and MDMT: RO+PA are similar to those with the X-axis.

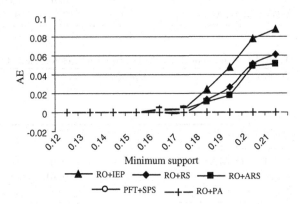

Fig. 4.4 AE versus α for experiments using database B

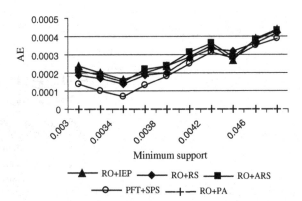

Fig. 4.5 AE versus α for experiments using database $R1$

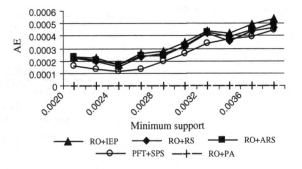

Fig. 4.6 AE versus α for experiments using database $R2$

4.8 Conclusions

In this chapter, we have discussed several generalized as well as specialized multi-database mining techniques. For the particular problem at hand, one technique could be more suitable than the others. However, we cannot claim that there is a single method of universal nature which outperforms all other techniques. Instead, a choice of the method has to be problem-driven. We have formalized the idea of multi-database mining using local pattern analysis by considering an underlying two-step process. We have also presented the pipelined feedback model which is particularly suitable for mining multiple large databases. It improves significantly the accuracy of mining multiple databases as compared to an existing technique that scans each database only once. The pipelined feedback model could also be used for mining a large database by dividing it into a series of sub-databases. Experimental results obtained with the use of the MDMT: PFT+SPS are promising and underline the usefulness of the method studied here.

References

Adhikari A (2013) Clustering local frequency items in multiple databases. Inf Sci 237:221–241

Adhikari A (2012) Synthesizing global exceptional patterns in different data sources. J Intelli Syst 21(3):293–323

Adhikari A, Rao PR (2008a) Synthesizing heavy association rules from different real data sources. Pattern Recogn Lett 29(1):59–71

Adhikari A, Rao PR (2008b) Efficient clustering of databases induced by local patterns. Decis Support Syst 44(4):925–943

Adhikari A, Ramachandrarao P, Pedrycz W (2011) Study of select items in different data sources by grouping. Knowl Inf Syst 27(1):23–43

Adhikari A, Ramachandrarao P, Prasad B, Adhikari J (2010) Mining multiple large data sources. Int Arab J Inf Technol 7(2):241–249

Adhikari J, Rao PR, Adhikari A (2009) Clustering items in different data sources induced by stability. Int Arab J Inf Technol 6(4):66–74

Agrawal R, Srikant R (1994) Fast algorithms for mining association rules. In: Proceedings of international conference on very large data bases, pp 487–499

Babcock B, Chaudhury S, Das G (2003) Dynamic sample selection for approximate query processing. In: Proceedings of ACM SIGMOD conference management of data, pp 539–550

Coenen F, Leng P, Ahmed S (2004) Data structure for association rule mining: T-trees and P-trees. IEEE Trans Knowl Data Eng 16(6):774–778

Frequent Itemset Mining Dataset Repository (2004) http://fimi.cs.helsinki.fi/data

Han J, Pei J, Yiwen Y (2000) Mining frequent patterns without candidate generation. In: Proceedings of ACM SIGMOD conference on management of data, pp 1–12

Savasere A, Omiecinski E, Navathe S (1995) An efficient algorithm for mining association rules in large databases. In: Proceedings of the 21st international conference on very large data bases, pp 432–443

Wu X, Zhang S (2003) Synthesizing high-frequency rules from different data sources. IEEE Trans Knowl Data Eng 14(2):353–367

Wu X, Zhang C, Zhang S (2005) Database classification for multi-database mining. Inf Syst 30(1):71–88

Zhang C, Liu M, Nie W, Zhang S (2004a) Identifying global exceptional patterns in multi-database mining. IEEE Comput Intell Bull 3(1):19–24
Zhang S, Wu X, Zhang C (2003) Multi-database mining. IEEE Comput Intell Bull 2(1):5–13
Zhang S, You X, Jin Z, Wu X (2009) Mining globally interesting patterns from multiple databases using kernel estimation. Expert Syst Appl 36(8):10863–10869
Zhang S, Zhang C, Yu JX (2004b) An efficient strategy for mining exceptions in multi-databases. Inf Sci 165(1–2):1–20

Chapter 5
Clustering Local Frequency Items in Multiple Data Sources

Frequent items could be considered as a generic type of pattern in a database. In the context of multiple data sources most of the global patterns are based on local frequency items. A multi-branch company transacting from different branches often needs to extract global patterns from data distributed over the branches. Global decisions could be made effectively using such patterns. Thus it becomes important to cluster local frequency items in multiple databases. In this chapter an overview of the existing measures of association is presented. For the purpose of selecting the suitable technique of mining multiple databases we have surveyed the existing multi-database mining techniques. A study on the related clustering techniques is also covered here. We present the notion of high frequency itemsets, and an algorithm for synthesizing the supports of such itemsets is designed. The existing clustering technique might cluster a set of items at a low level since it estimates association among items in an itemset with low accuracy, and thus a new algorithm for clustering local frequency items is proposed. Due to the suitability of measure of association A_2, on its basis, association among items in a high frequency itemset is synthesized. The soundness of the clustering technique has been shown. Numerous experiments are conducted using five datasets, and the results concerning different aspects of the proposed problem are presented in the experimental section. The effectiveness of the proposed clustering technique is more visible in dense databases.

5.1 Introduction

Many multi-branch companies collect a huge amount of data through different branches and the local transactions are stored locally. Banks, shopping malls, and insurance companies are some examples of multi-branch companies that deal with multiple databases. In the context of multiple databases, many global patterns such as high frequency association rule (Wu and Zhang 2003), heavy association rule (Adhikari and Rao 2008d), and exceptional pattern (Adhikari 2012) are based on local frequency items in multiple databases. Thus, knowledge discovery using

A. Adhikari et al., *Data Analysis and Pattern Recognition in Multiple Databases*, Intelligent Systems Reference Library 61, DOI: 10.1007/978-3-319-03410-2_5, © Springer International Publishing Switzerland 2014

local frequency items becomes an important issue. The goal of this chapter is to present a technique for clustering local frequency items in multiple databases.

For the purpose of continuity of the presentation, we start with a few definitions. An *itemset* is a collection of items in a database. Each itemset in a database is associated with a statistical measure called *support* (Agrawal et al. 1993). The support of an itemset X in database D is expressed as the fraction of transactions in D containing X, denoted by $S(X, D)$. In general, let $S(E, D)$ be the support of a Boolean expression E defined on the transactions in database D. An itemset X is called *frequent* in D if $S(X, D) \geq \alpha$, where α is user-defined level of *minimum support*. If X is frequent then $Y \subseteq X$ is also frequent, since $S(Y, D) \geq S(X, D)$, for $Y \neq \phi$. Thus, each item of a frequent itemset is also frequent. Items in a frequent itemset could be considered as a basic type of patterns in a transactional database.

The collection of frequent itemsets determines major characteristics of a database. Many interesting algorithms (Agrawal and Srikant 1994; Han et al. 2000; Savasere et al. 1995) have been proposed for mining frequent itemsets in a database. Thus, there are many implementations for extracting frequent itemsets from a database (Frequent itemset mining implementations repository). Itemset patterns influence heavily the current KDD research. We observe such influence in the following way. Firstly, many algorithms as mentioned above, have been reported about mining frequent itemsets in a database. Secondly, many patterns are based on frequent itemsets in a database. They could be called as derived patterns in a database. For example, positive association rule (Agrawal et al. 1993), high frequency association rule (Wu and Zhang 2003) and conditional pattern (Adhikari and Rao 2008b) are examples of some derived patterns. Considerable number of studies have been reported on mining/synthesizing derived patterns in a database (Adhikari and Rao 2008d; Zhang et al. 2004a, b; Adhikari 2012). Finally, solutions to many problems could be based on the analysis of patterns in a database (Adhikari and Rao 2008c; Wu et al. 2005). Such applications process patterns in a database for the purpose of making some decisions. Frequent items are the components of many interesting patterns. Thus, the analysis and synthesis of local frequency items is an interesting as well as an important issue. They are used to construct the global patterns in multiple databases. Thus, clustering local frequency items in multiple databases is an important knowledge for a multi-branch company. Many corporate decisions could be taken effectively by incorporating knowledge inherent in data across the branches. An effective management of multiple large databases becomes a challenging issue (Adhikari et al. 2010a). In the next section we study the existing techniques of mining multiple large databases. Based on their suitability and performances, we choose the best among the available techniques for mining multiple databases.

This chapter is organized as follows. Section 5.2 elaborates on a study of the existing measures of association, techniques for mining multiple databases, and related clustering algorithms. The proposed problem is presented in Sect. 5.3. In Sect. 5.4, an algorithm for synthesizing supports of high frequency itemsets is designed. The algorithm also synthesizes association among items in a high frequency itemset of size greater than one. In Sect. 5.5, we design an algorithm for clustering local frequency items in multiple databases. Finally, experimental results are presented in Sect. 5.6 to show the effectiveness of the proposed clustering technique.

5.2 Related Work

Consider a multi-branch company that operates from n branches. Let D_i be the database corresponding to the ith branch, $i = 1, 2, \ldots, n$. Also, let D be the union of these databases. Our clustering procedure is based on itemset patterns in multiple databases. In this context, one needs a technique for mining itemset patterns in multiple databases. Afterwards, association among items in an itemset is captured using a measure of association. Finally, a clustering algorithm to cluster local frequency items in multiple databases is designed. We categorize work related to this issue into three areas viz. measures of association, techniques for mining multiple databases, and related clustering algorithms.

5.2.1 Measures of Association

The analysis of relationships among variables is a fundamental task being at the heart of many data mining problems. For instance, association rules find relationships between sets of items in a database of transactions. Such rules express buying patterns of customers e.g., finding how the presence of one item affects the presence of another, and so forth. A measure of association gives a numerical estimate of statistical dependence among a set of items. Highly associated items are likely to be purchased together. In other words, items of itemset X are highly associated, if one of the items of X is purchased then the remaining items of X are also likely to be purchased in the same transaction.

Tan et al. (2003) have described several key properties of twenty one interestingness measures proposed in statistics, machine learning and data mining literature. It might be required to examine the properties in order to select right interestingness measure for a given application domain. Hershberger and Fisher (2005) discuss some measures of association proposed in statistics. Measures of association could be categorized into two groups. Some measures deal with a set of objects, or could be generalized to deal with a set of objects. On the other hand, remaining measures could not be generalized. Confidence (Agrawal et al. 1993; conviction (Brin et al. 1997) are examples of the second category of measures. On the other hand, measures such as Jaccard (Tan et al. 2003) could be generalized to find association among a set of items in a database. Most of the existing measures are based on a 2×2 contingency table. Thus, these measures might not be suitable for measuring association among a set of items.

Agrawal et al. (1993) have proposed support measure in the context of finding association rules in a database. To find support of an itemset, it requires counting frequency of the itemset in the given database. An itemset in a transaction could be a source of association among items in the itemset. But, support of an itemset does not consider frequencies of it subsets. As a result, the support of an itemset might not be a good measure of association among items in an itemset.

Piatetsky-Shapiro (1991) has proposed leverage measure in the context of mining strong rules in a database. Aggarwal and Yu (1998) have proposed a notion of a collective strength of an itemset. Collective strength is based on the concept of violation of an itemset. An itemset X is said to be in violation of a transaction, if some items of X are present in the transaction and others are not. Collective strength of an itemset X has been defined as follows.

$$C(X) = \frac{1 - v(X)}{1 - E(v(X))} \times \frac{E(v(X))}{v(X)}, \quad \text{where} \tag{5.1}$$

- $v(X)$ is the violation rate of itemset X. It is the fraction of transactions in violation of itemset X.
- $E(v(X))$ is the expected violation rate of itemset X.

The major concern regarding computation of $C(X)$ is that the computation of $E(v(X))$ is based on statistical independence of items of X.

Cosine (Han and Kamber 2001) and correlation (Han and Kamber 2001) are used to measure association between two objects. These measures might not be suitable as a measure of association among items of an itemset. Confidence and conviction are used to measure strength of association between itemsets in some sense. These measures might not be useful in the current context, since we are interested in capturing association among items of an itemset.

Zhou and Xiong (2009) have proposed to find confounding effects attributable to local associations efficiently. Authors derived an upper bound by a necessary condition of confounders, which can help us prune the search space and efficiently identify confounders. Duan and Street (2009) explored high-dimensional correlation in two ways. Initially, authors expanded the set of desirable properties for correlation measures and study the advantages and disadvantages of various measures. Then, they proposed an MFCI framework to decouple the correlation measure from the need for efficient search.

5.2.2 Multi-Database Mining Techniques

Mining multiple large databases seems to be a different problem than mining a database (Adhikari et al. 2010a). One of the main challenges is that the collection of all branch databases might be very large. Traditional data mining techniques

(Agrawal and Srikant 1994; Han et al. 2000) seem to be not suitable for mining multiple large databases. In this situation, one could employ local pattern analysis (Zhang et al. 2003) to deal with multiple large databases. In this technique each branch is required to forward local patterns, instead of original databases, to the central office for synthesis and analysis of local patterns. Local pattern analysis might return approximate global patterns.

For the purpose of mining multiple databases, one could apply *partition algorithm* proposed by Savasere et al. (1995). The algorithm was designed to mine a very large database by partitioning. The algorithm works as follows. It scans the database twice. The database is divided into disjoint partitions, where each partition is small enough to fit in memory. In the first scan, the algorithm reads each partition and computes locally frequent itemsets in each partition using apriori algorithm (Agrawal and Srikant 1994). In the second scan, the algorithm counts the supports of all locally frequent itemsets toward the complete database. In this case, each local database could be considered as a partition. Though partition algorithm mines frequent itemsets accurately, it is an expensive solution to mining multiple large databases, since each database is required to be mined twice.

Zhang et al. (2004a, b) have proposed algorithm *IdentifyExPattern* for identifying global exceptional patterns in multi-databases. Every local database is mined separately using *random order* for synthesizing global exceptional patterns. In algorithm *IdentifyExPattern*, a pattern in a local database is assumed as zero, if it does not get reported. Let $S_a(p, DB)$ and $S_s(p, DB)$ be the actual (i.e., apriori) support and synthesized support of pattern p in database DB. Support of pattern p in D has been synthesized in the form

$$S_s(p, D) = \frac{1}{num(p)} \sum_{i=1}^{num(p)} \frac{S_a(p, D_i) - \alpha}{1 - \alpha} \qquad (5.2)$$

where $num(p)$ is the number of databases that report pattern p at a given minimum support level α.

For synthesizing association rules in multiple real databases *Association-Rule-Synthesis* algorithm has been proposed in (Adhikari and Rao 2008d). For real databases, the trend of the customers' behaviour exhibited in one database is usually present in other databases. In particular, a frequent itemset in one database is usually present in some transactions of other databases even if it does not get extracted. The estimation procedure captures such trend, and estimates the support of a missing association rule. Without any loss of generality, let the itemset X be extracted from first m databases, $1 \leq m \leq n$. Then trend of X in first m databases could be expressed as follows:

$$trend^{1,m}(X|\alpha) = \frac{1}{\sum_{i=1}^{m} |D_i|} \times \sum_{i=1}^{m} [S_a(X, D_i) \times |D_i|] \qquad (5.3)$$

One could use the trend of X in first m databases for synthesizing support of X in D. An estimate of support of X in each of the remaining databases is obtained by $\alpha \times trend^{1,n}(X \mid \alpha)$. Thus, the synthesized support of X is computed as follows:

$$S_s(X, D) = \frac{trend^{1,m}(X|\alpha)}{\sum_{i=1}^{n} |D_i|} \times \left[(1 - \alpha) \times \sum_{i=1}^{m} |D_i| + \alpha \times \sum_{i=1}^{n} |D_i| \right] \qquad (5.4)$$

For synthesizing high frequency association rules in multiple databases, Wu and Zhang (2003) have proposed *RuleSynthesizing* algorithm. Based on the association rules in different databases, the authors have estimated weights of different databases. Let w_i be the weight of ith database, $i = 1, 2, ..., n$. Without any loss of generality, let the association rule r be extracted from first m databases, $1 \le m \le n$. Then $S_a(r, D_i)$ has been assumed as 0, for $i = m + 1, m + 2, ..., n$. Then support of r in D has been synthesized in the following form.

$$S_s(r, D) = w_1 \times S_a(r, D_1) + \cdots + w_m \times S_a(r, D_m) \qquad (5.5)$$

He et al. (2010) studied the problem of rule synthesizing from multiple related databases where items representing the databases may be different, and the databases may not be relevant, or similar to each other. Authors argued that, for such multi-related databases, simple rule synthesizing without a detailed understanding of the databases is not able to reveal meaningful patterns inside the data collections. Consequently, authors proposed a two-step clustering on the databases at both item and rule levels such that the databases in the final clusters contain both similar items and similar rules. A weighted rule synthesizing method then has been applied to each such cluster to generate final rules.

Adhikari et al. (2011b) have introduced a new pattern, called notch, of an item in time-stamped databases. Based on this pattern, authors have proposed two special kinds of notch, called generalized notch and iceberg notch, in time-stamped databases. Also, the authors have identified an application of generalized notch, and designed an algorithm for mining interesting icebergs in time-stamped databases.

The variation of sales of an item over time is an important issue. Thus, the notion of stability of an item is introduced in (Adhikari et al. 2009). Stable items are useful in making many strategic decisions for a company. Based on the degree of stability of an item, an algorithm has been designed for clustering items in different data sources. A notion of best cluster by considering average degree of variation of a class has been proposed. Also, an alternative algorithm has been designed to find best cluster among items in different data sources.

Liu et al. (2001) have proposed multi-database mining technique that searches only the relevant databases. Identifying relevant databases is based on selecting the relevant tables (relations) that contain specific, reliable and statistically significant information pertaining to the query. Adhikari et al. (2010a), Zhang et al. (2004a, b) studied various strategies for mining multiple databases. Existing parallel mining techniques could also be used to deal with multiple large databases (Agrawal and Shafer 1996; Chattratichat et al. 1997; Cheung et al. 1996).

Many large organizations have multiple large databases as they transact from multiple branches. Numerous decisions are based on a set of specific items called the select items. Thus, the analysis of select items in multiple databases is an important issue. For the purpose of studying select items in multiple databases, one may need true global patterns of select items. Thus, we have proposed a model of mining global patterns of select items from multiple databases (Adhikari et al. 2011a). A measure of overall association between two items in a database is also proposed. An algorithm is designed based on the proposed measure for the purpose of grouping the frequent items in multiple databases.

Data collected for collaborative filtering (CF) might be cross-distributed between two online vendors, even competing companies. Such corporations might want to integrate their data to provide more precise and reliable recommendations. However, due to privacy, legal, and financial concerns, they do not desire to disclose their private data to each other. If privacy-preserving measures are introduced, they might decide to generate predictions based on their distributed data collaboratively. Yakut and Polat (2012) have investigated how to offer hybrid CF-based referrals with decent accuracy on cross-distributed data (CDD) between two e-commerce sites while maintaining their privacy.

5.2.3 Clustering Techniques

Zhang et al. (1997) have proposed an efficient and scalable data clustering method, called BIRCH, based on a new in-memory data structure called CF-tree. Estivill-Castro and Yang (2004) have proposed an algorithm that remains efficient, generally applicable, multi-dimensional but is more robust to noise and outliers. Jain et al. (1999) have presented an overview of pattern clustering methods from a statistical pattern recognition perspective, with a goal of providing useful advice and references to fundamental concepts accessible to the broad community of clustering practitioners. In this chapter, we would like to design a clustering algorithm based on local patterns. Thus, the above algorithms might not be suitable in this situation.

Ali et al. (1997) have proposed a partial classification technique using association rules. The clustering of frequent items using local association rules might not be a good idea. The number of frequent itemsets obtained from a set of association rules might be much less than the number of frequent itemsets extracted using apriori algorithm. Thus, efficiency of the clustering process would become low. A measure of similarity between two databases, $simi_2$, is proposed in (Adhikari and Rao 2008c) and an algorithm is designed to cluster databases using measure $simi_2$.

Chen et al. (2012) have proposed a new algorithm to cluster multiple and parallel data streams using spectral component similarity analysis. The algorithm performs auto-regressive modeling to measure the lag correlation between the data streams and uses it as the distance metric for clustering. The algorithm uses a sliding window model to continuously report the most recent clustering results and to dynamically adjust the number of clusters. Malinen and Fränti (2012) have shown that clustering is also an optimization problem for an analytic function. The mean squared error or the squared error can be expressed as an analytic function. With an analytic function one can get benefit from the existence of standard optimization methods. Initially the gradient of this function is calculated and then the descent method is used to minimize the function.

Identifying clusters of arbitrary shapes remains a challenge in the field of data clustering. Lee and Ólafsson (2011) have proposed a new measure of cluster quality based on minimizing the penalty of disconnection between objects that would be ideally clustered together. This disconnectivity is based on analysis of nearest neighbors and the principle that an object should be in the same cluster as its nearest neighbors. An algorithm called MinDisconnect is proposed that heuristically minimizes disconnectivity and numerical results are presented that indicate that the new algorithm can effectively identify clusters of complex shapes and is robust in finding clusters of arbitrary shapes.

Mampaey and Vreeken (2013) have proposed an approach to build a summary by clustering attributes that strongly correlate, and uses the minimum description length principle to identify the best clustering, without requiring a distance measure between attributes.

5.3 Problem Statement

Based on the number of data sources, patterns in multiple databases could be classified into three categories. They are local patterns, global patterns, and patterns that are neither local nor global. Local patterns are used for local data analyses. On the other hand, global patterns are based on all the databases. They are used for global data analyses (Adhikari and Rao 2008d; Wu and Zhang 2003).

In the context of mining multiple databases using local pattern analysis, some itemsets are reported from many databases. They could be termed as high frequency itemsets. In this chapter, we synthesize high frequency itemsets based on local itemsets. We measure association among items in high frequency itemsets. Then we cluster frequent items based on associations obtained among items in high frequency itemsets. Highly associated items could be put in the same class. The motivation of proposed clustering technique is given as follows.

An approach for grouping databases, called existing technique, has been proposed in (Adhikari and Rao 2008c; Wu et al. 2005). The grouping technique could be described as follows. It finds association between every pair of items (databases). A set of m arbitrary items forms a class, if the $^mC_2$ pair-wise associations corresponding to $^mC_2$ pairs of items are close. The level of association among the items in this class is assumed as the minimum of $^mC_2$ associations. If the number of items in a class is higher than two, then we observe that this technique might fail to estimate correctly the association among the items in a group. Then accuracy of the entire clustering process becomes low. The proposed clustering technique

Table 5.1 Symbols/abbreviations used in the study

Symbol/abbreviation	Meaning/full form
α	User-defined minimum support level
$FI(D \mid \alpha)$	Set of frequent items in database D at a given α
$FIS(D \mid \alpha)$	Set of frequent itemsets in database D at a given α
$FI(1, n, \alpha)$	$\cup_{i=1}^{n} FI(D_i \mid \alpha)$
$FIS(1, n, \alpha)$	$\cup_{i=1}^{n} FIS(D_i \mid \alpha)$
γ	Minimum threshold of number of extractions of an itemset
n	Number of databases
HFIS	High frequency itemset
SHFIS	Synthesized HFIS
AIS	Two dimensional array such that $AIS(i)$ is the array of itemsets extracted from D_i
IS	Set of all itemsets in n databases
$S(X, D_1)$	Support of itemset in databases D_1
SA	Synthesized association
SD	Synthesized dispersion
δ	Level of association

follows a different approach and it clusters local frequency items at a higher level of similarity as compared to existing technique.

Before stating our problem formally, we incorporate all the notations in the above table for easy reference (Table 5.1).

Each item in $FI(D_i|\alpha)$ is a local frequency item, $i = 1, 2, 3, \ldots$. We apply a measure of association and a multi-database mining technique for the purpose of clustering local frequency items in multiple databases. The proposed problem could be stated as follows.

There are n different databases D_i, $i = 1, 2, \ldots, n$. Find the best non-trivial partition (if it exists) of $FI(1, n, \alpha)$ induced by $FIS(1, n, \alpha)$.

A partition (Liu 1985) is a specific type of clustering. A formal definition of a non-trivial partition is given in Sect. 5.5.

5.4 Synthesizing Support of an Itemset

We have designed technique PFT for mining multiple large databases (Adhikari et al. 2010b). It improves the quality of global pattern as compared to existing techniques. We shall use this technique here for mining multiple large databases. Let the itemset X be extracted from k out of n databases using PFT, $0 < k \leq n$. Let γ be the minimum threshold of number of extractions of an itemset, $0 < \gamma \leq 1$. We would be interested in an itemset if it has been extracted from at least $n \times \gamma$ databases. Such itemsets are called *high frequency itemsets* (*HFIS*s). If an itemset X has high frequency then another itemset $Y \subseteq X$ also has high frequency, for $Y \neq \phi$. We define a high frequency itemset as follows.

Definition 5.1 Let there be n databases. Let X be an itemset extracted from k $(\leq n)$ databases. Then X has high frequency if $\frac{k}{n} \geq \gamma$, where γ is the minimum threshold of the number of extractions of an itemset, $0 < \gamma \leq 1$. •

A high frequency itemset might not be frequent in all the databases. A high frequency itemset may not be extracted naturally when we apply PFT. After applying PFT, we synthesize supports of *HFIS*s. In (Adhikari and Rao 2008a), A_2 measure has been proposed to find association among items in a database. A detailed study on existing measures of association is completed for computing association among items in a database. It has been shown that the measure A_2 is effective in computing association among items in an itemset. In Example 5.1, we illustrate the procedure for synthesizing support of a *HFIS*.

Example 5.1 Consider a multi-branch company that has four branches. Let D_i be the database corresponding to ith branch, $i = 1, 2, 3, 4$. The branch databases are given as follows. $D_1 = \{\{a, b\}, \{a, b, c\}, \{a, b, c, d\}, \{c, d, e\}, \{c, d, f\}, \{c, d, i\}\}$; $D_2 = \{\{a, b\}, \{a, b, g\}, \{g\}\}$; $D_3 = \{\{a, b, d\}, \{a, c, d\}, \{c, d\}\}$, $D_4 = \{\{a\}, \{a, b, c\}, \{c, d\}, \{c, d, i\}\}$. Assume that $\alpha = 0.4$, and $\gamma = 0.6$. Let $X(\eta)$ denote the fact that the itemset X has support η in the database. We sort databases in a non-increasing order on database size (expressed in bytes). The sorted databases are given as follows: D_1, D_4, D_3, D_2. By applying PFT, we obtain the following itemsets in different local databases:

$LPB(D_1, \alpha) = \{\{a\}(0.5), \{b\}(0.5), \{c\}(0.833), \{d\}(0.667), \{a, b\}(0.5), \{c, d\}(0.667)\}$, $LPB(D_4, \alpha) = \{\{a\}(0.667), \{b\}(0.25), \{c\}(0.75), \{d\}(0.25), \{a, b\}(0.333), \{c, d\}(0.667)\}$, $LPB(D_3, \alpha) = \{\{a\}(0.667), \{b\}(0.333), \{c\}(0.667), \{d\}(1.0), \{a, b\}(0.333), \{c, d\}(0.667)\}$, and $LPB(D_2, \alpha) = \{\{a\}(0.667), \{b\}(0.667), \{c\}(0.0), \{d\}(0.0), \{a, b\}(0.667), \{c, d\}(0.0), \{g\}(0.667)\}$.

Let D be the union of D_1, D_2, D_3 and D_4. Synthesized *HFIS*s in D are given as follows: $SHFIS(D, 0.4, 0.6) = \{\{a\}(0.563), \{b\}(0.438), \{c\}(0.563), \{d\}(0.563), \{a, b\}(0.438), \{c, d\}(0.5)\}$. •

Since we apply PFT for mining multiple databases, the *RuleSynthesizing* algorithm (Wu and Zhang 2003) is not applicable here. Moreover, PFT is a direct approach for estimating support of a high frequency itemset.

The collection of *SHFIS*s of size greater than one forms the basis of the proposed clustering technique. We present below an algorithm to form synthesized association among items in each *SHFIS* of size greater than one. Let N be the number of itemsets in the given n databases. Let *AIS* be a two dimensional array such that $AIS(i)$ is the array of itemsets extracted from D_i, $i = 1, 2, ..., n$. Also, let *IS* be the set of all itemsets in n databases. An itemset could be described by the following attributes: *itemset*, *supp*, and *did*. Here, *itemset*, *supp* and *did* denote the itemset, support and database identification of an itemset, respectively. All the synthesized itemsets are kept in *SIS*, an array of synthesized itemset. Each synthesized itemsets has the following attributes: *itemset*, *ss*, and *sa*. Here, *ss* and *sa* denote the synthesized support and synthesized association of the itemset, respectively. In the following algorithm (Adhikari 2013), we synthesize association among items of each *SHFIS*.

Algorithm 5.1 Synthesize association among items of each *SHFIS* of size greater than one.

procedure *SynthesizingAssociation* (*n*, *AIS*, *size*, γ)

Input:

n: number of databases

AIS: two-dimensional array of itemsets extracted during mining multiple databases

size: array of number of transactions in input databases

γ: threshold for minimum number of extractions of an itemset

Output:

Synthesized association among items of each *SHFIS*

```
01:   store all the local itemsets into array IS;
02:   sort itemsets of IS based on itemset attribute;
03:   add sizes of all branch databases into variable totalTransactions;
04:   let nSynItemSets = 0; let i = 1;
05:   if ( i ≤ |IS| ) then
06:     let j = i; let count = 0;
07:     while (j ≤ i + n) do
08:       if (IS(j).itemset = IS(i).itemset) then
09:         process support of IS(i);
10:         increase count by 1; increase j by 1;
11:       else go to line 14;
12:       end if
13:     end while {07}
14:     synSupp = suppₛ(IS(i).itemset, D);
15:     if (count / n ≥ γ) then
16:       SIS(nSynItemSets). ss  = synSupp;
17:       SIS(nSynItemSets). itemset  = IS(i).itemset;
18:     end if
19:     let i = j;
20:     increase nSynItemSets by 1;
21:     go to line 5;
22:   end if {05}
23:   for j = 1 to nSynItemSets do
24:     if (|SIS(nSynItemSets). itemset| ≥ 2) then
25:       SIS(nSynItemSets). sa = A₂(SIS(nSynItemSets). itemset, D);
26:     end if
27:   end for {23}
end procedure
```

We sort itemsets of *IS*, so that the same itemset extracted from different data sources comes consecutive. It helps processing itemsets easier. We find total number of transactions in different databases into variable *totalTransactions*.

The variables *nSynItemSets* and *i* keep track of the number synthesized itemsets and the current itemset of *IS*, respectively. The algorithm segment in lines 5–22 is repeated *N* times. An itemset gets processed in an iteration. An itemset occurs maximum *n* times, since there are *n* databases. Thus, the while-loop in lines 7–13 repeats maximum *n* times. The variable *count* keeps track of number of times an itemset gets extracted. Based on variable *count* we could determine whether an itemset has high frequency. If an itemset is highly frequent then we store the details into array *SIS*, and increase *nSynItemSets* by 1. We update variable *i* by *j* for processing the next itemset. We go back to line 5 for processing the next itemset. Using lines 23–27, we calculate synthesized association (Adhikari and Rao 2008a), for each synthesized itemset of size greater than one. In the next paragraph, we determine the time complexity of above algorithm.

Line 1 takes $O(N)$ time, since there are *N* itemsets in *n* databases. Line 2 takes $O(N \times \log(N))$ time to sort *N* itemsets. Line 3 takes $O(n)$ time, since there are *n* databases. The *while* loop at line 7 repeats maximum *n* times. The *if* statement at line 5 repeats *N* times. Thus, time complexity of program segment in lines 5–22 is $O(n \times N)$. Line 23 takes $O(N)$ time. Let the average size of an itemset be *p*. The time complexity for searching an itemset in *IS* is $O(N)$. The time-complexity for computing association of an itemset is $O(N \times p^2)$. Thus, the time complexity of program segment in lines 23–27 is $O(N^2 \times p^2)$ time. Therefore, the time complexity of procedure *SynthesizingAssociation* is equal to *maximum* $\{O(N^2 \times p^2),$ $O(N \times \log(N)), O(n \times N)\} = O(N^2 \times p^2)$, since $N > \log(N)$ and $N > n$.

5.5 Clustering Local Frequency Items

Local frequency items are the basic components of many important patterns in multiple databases. The main objective of this study is to cluster local frequency items. The existing techniques (Wu et al. 2005; Adhikari and Rao 2008c) of clustering multiple databases work as follows. A measure of similarity between two databases is proposed. Let there be *m* databases to be clustered. Then the similarities for $^mC_2$ pairs of databases are computed. An algorithm is designed based on the measure of similarity. Then based on a level of similarity, the databases are clustered into different classes.

For the purpose of clustering databases, Wu et al. (2005) have proposed the following measure of similarity between two databases.

$$sim_1(D_1, D_2) = \frac{|I(D_1) \cap I(D_2)|}{|I(D_1) \cup I(D_2)|} \tag{5.6}$$

Here $I(D_i)$ is the set of items in the database D_i, $i = 1, 2$. Thus, sim_1 is the ratio of number of items common to databases and the total number of items in the two databases. Measure sim_1 could also be used to find similarity between two items in a database. Also, we observe that the measure sim_1 could be obtained from A_2

(Adhikari and Rao 2008a). In other words, the measure sim_1 is a special case of measure A_2. In the following example, we show that association among items of an itemset could not be determined correctly using associations of all possible subsets of size 2. In particular, association among items of $\{a, b, c\}$ could not be correctly estimated by the associations between the items in $\{a, b\}$, $\{a, c\}$, and $\{b, c\}$. We explain this issue by using Example 5.2.

Example 5.2 Let $D_5 = \{\{a, b, c, d\}, \{a, b, c, e\}, \{a, b, d\}, \{a, e, f\}, \{b, c, e\}, \{d, e, g\}, \{d, f, g\}, \{e, f, g\}, \{e, f, h\}, \{g, h, i\}\}$. Also, let α be 0.2. The supports of relevant frequent itemsets are given as follows. $S(\{a\}, D_5) = 0.4$, $S(\{b\}, D_5) = 0.4$, $S(\{c\}, D_5) = 0.3$, $S(\{a, b\}, D_5) = 0.3$, $S(\{a, c\}, D_5) = 0.2$, $S(\{b, c\}, D_5) = 0.3$, $S(\{a, b, c\}, D_5) = 0.2$. Now, $sim_1(\{a, b\}, D_5) = 0.6$, $sim_1(\{a, c\}, D_5) = 0.4$, $sim_1(\{b, c\}, D_5) = 0.75$. Using sim_1, the items a, b, and c could be put in the same class at the level of similarity 0.4, i.e., $minimum\{0.6, 0.4, 0.75\}$. Using A_2, we have $A_2(\{a, b, c\}, D_5) = 0.67$. Thus, the items a, b, and c could be put in the same class at the level 0.67. We observe that the subset of transactions $\{\{a, b, c, d\}, \{a, b, c, e\}, \{a, b, d\}, \{a, e, f\}, \{b, c, e\}\}$ of D_5 that results the amount of association among a, b, and c. Three out of five transactions contain at least two items of $\{a, b, c\}$. Two out of five transactions contain all the items of $\{a, b, c\}$. More the number of items of $\{a, b, c\}$ occur together, higher is the association among items of $\{a, b, c\}$. Thus, we observe that the level of association among the items of $\{a, b, c\}$ is close to 0.67 rather than 0.4. Thus, we fail to measure association correctly among the items of $\{a, b, c\}$ based on the similarities between items of $\{a, b\}$, $\{a, c\}$, and $\{b, c\}$. •

Example 5.2 shows that the existing clustering technique (Wu et al. 2005; Adhikari and Rao 2008c) might cluster a set of items at a low accuracy. This is due to the fact that we are unable to estimate the association among the items in an itemset. If we cluster the items at a given similarity level using the existing technique, then items in a class become more similar than the given level of similarity. Thus, the degree of similarity among items in a class is not true representative of the similarity level of a clustering algorithm. The proposed clustering algorithm improves the accuracy of clustering, and it is more visible in dense databases. We arrive at the following observations.

Observation 5.1 Let $X = \{x_1, x_2, ..., x_m\}$ be an itemset in database D. The *BestClassification* algorithm (Wu et al. 2005) might put items of X in a class at the level *minimum* $\{sim_1(x_i, x_j, D) : 1 \leq i < j \leq m\}$.

Observation 5.2 The proposed clustering technique might put items of X in a class at the level of association $A_2(\{x_1, x_2, ..., x_m\}, D)$.

A clustering of items results in a set of classes of items. A class of frequent items over $FI(1, n, \alpha)$ could be defined as follows.

Definition 5.1 A class $class^\delta$ formed at a level of association δ under the measure of association A_2 over $FI(1, n, \alpha)$ in database D is defined as $X \subseteq FI(1, n, \alpha)$ such that $A_2(X, D) \geq \delta$, and one of the following conditions is satisfied: (1) $X \in$ $SHFIS(1, n, \alpha, \gamma)$, for $|X| \geq 2$, (2) $X \subseteq FI(1, n, \alpha)$, for $|X| = 1$. •

Definition 5.2 enables us to obtain and define a clustering of frequent items over $FI(1, n, \alpha)$ as follows.

Definition 5.2 Let π^δ be a clustering of local frequency items over $FI(1, n, \alpha)$ at level of association δ under the measure of association A_2. Then, $\pi^\delta = \{X : X$ is a class of type $class^\delta$ over $FI(1, n, \alpha)\}$. •

We symbolize the ith class of π^δ as $CL_i^{\delta,\alpha}$, $i = 1, 2, ..., |\pi^\delta|$. A clustering might not include all the local frequency items. We might be interested in clustering all local frequency items. A complete clustering of local frequency items over $FI(1, n, \alpha)$ is defined as follows.

Definition 5.3 A clustering $\Pi^\delta = \left\{ CL_1^{\delta,\alpha}, CL_2^{\delta,\alpha}, ..., CL_m^{\delta,\alpha} \right\}$ is complete, if $\cup_{i=1}^m$ $CL_i^{\delta,\alpha} = FI(1, n, \alpha)$, where $CL_i^{\delta,\alpha}$ is a class of type $class^\delta$ over $FI(1, n, \alpha)$, $i = 1$, $2, ..., m$. •

Two classes in a clustering might not be mutually exclusive. We might be interested in determining a mutually exclusive clustering. A mutually exclusive clustering over $FI(1, n, \alpha)$ could be defined as follows.

Definition 5.4 A clustering $\Pi^\delta = \left\{ CL_1^{\delta,\alpha}, CL_2^{\delta,\alpha}, ..., CL_m^{\delta,\alpha} \right\}$ is mutually exclusive if $CL_i^{\delta,\alpha} \cap CL_j^{\delta,\alpha} = \varphi$, $CL_i^{\delta,\alpha}$ and $CL_j^{\delta,\alpha}$ are classes of type $class^\delta$ over $FI(1, n, \alpha)$, $i \neq j, i, j = 1, 2, ..., m$. •

One might be interested in finding out such a mutually exclusive and complete clustering. We are interested in finding the best non-trivial partition, if it exists, of local frequency items. First, we define a partition of local frequency items as follows.

Definition 5.5 A complete and mutually exclusive clustering is called a partition. •

A clustering is not necessarily to be a partition. In most of the cases, a trivial partition might not be of interest to us. We define a non-trivial partition of local frequency items as follows.

Definition 5.6 A partition π is non-trivial if $1 < |\pi| < n$, where n is the number of items in different databases. •

A partition is based on $SHFIS$s and associations among items in these itemsets. For this purpose, we need to synthesize association among items of every $SHFIS$ of size greater than one. We define a synthesized association among items of a $SHFIS$ as follows.

Definition 5.7 Let there be n different databases. Let $X \in SHFIS(D)$ such that $|X| \geq 2$. Synthesized association among the items of X, computed using A_2, is denoted by $SA(X, D| \alpha, \gamma)$. •

To determine goodness of a partition, we need to measure dispersion among items in a 2-item *SHFIS*. We define synthesized dispersion, *SD* of an itemset of size 2, as follows.

Definition 5.8 Let there be n different databases. Let $X \in SHFIS(D)$ such that $|X| = 2$. Synthesized dispersion *SD* between the items of X is given by

$$SD(X, D|\alpha, \gamma) = 1 - SA(X, D|\alpha, \gamma). \bullet \tag{5.7}$$

We calculate synthesized associations of all *SHFIS*s of size greater than one. In Example 5.3, we calculate synthesized associations of itemsets in *SHFIS(D)*.

Example 5.3 We continue here the discussion of Example 5.1. Synthesized associations among items of relevant *SHFIS*s are given as follows: $SA(\{a, b\}, D) = 0.78$, $SA(\{c, d\}, D) = 0.80$. We arrange *SHFIS*s of size greater than one in non-increasing order on synthesized association. The arranged *SHFIS*s are given as follows: $\{c, d\}, \{a, b\}$. Also, $FI((1, n, \alpha)) = \{a, b, c, d, g\}$. There exist two non-trivial partitions. They are given as follows: $\pi^{0.80} = \{\{a\}, \{b\}, \{g\}, \{c, d\}\}$, and $\pi^{0.78} = \{\{g\}, \{a, b\}, \{c, d\}\}$. $\bullet$

A local frequency item that is extracted only from a few databases forms a singleton class. In above partitions, g is an example of such an item.

5.5.1 Finding the Best Non-Trivial Partition

In Example 5.3, we observe the existence of two non-trivial partitions at the levels of association 0.80 and 0.78. We would like to find the best partition among the available non-trivial partitions. The best partition is based on the principle of maximizing the intra-class association and maximizing inter-class dispersion. Intra-class association and inter-class dispersion are defined as follows.

Definition 5.9 The *intra-class association* of partition π at the level of association δ under the measure of synthesized association *SA* is defined as follows.

$$intra\text{-}class\ association(\pi^{\delta}) = \sum_{C \in \pi, |C| \geq 2} SA(C|\alpha, \gamma). \bullet \tag{5.8}$$

In partition π, there may exist classes of size greater than or equal to 2. Intra-class association of π is obtained by adding synthesized associations of those classes.

Definition 5.10 The *inter-class dispersion* of partition π at the level of association δ under the measure of synthesized dispersion *SD* is defined as follows.

$$inter\text{-}class\ dispersion(\pi^\delta) = \sum_{C_p, C_q \in \pi; p \neq q} \sum_{a \in C_p, b \in C_q; \{a,b\} \in SHFIS} SD(\{a,b\}|\alpha,\gamma). \bullet$$

$$(5.9)$$

Between the two classes C_p and C_q in π, the synthesized dispersion is obtained by adding synthesized dispersion of $\{a, b\}$, where $a \in C_p$, and $b \in C_q$, $p \neq q$. Then inter-class dispersion of π is obtained by adding synthesized dispersion between items in C_p and C_q, where C_p, $C_q \in \pi$, $p \neq q$. We would like to define goodness measure of a partition for the purpose of finding the best partition among available non-trivial partitions. We define goodness measure of a partition as follows.

Definition 5.11

$$goodness(\pi^\delta) = intra\text{-}class\ association(\pi^\delta) + inter\text{-}class\ dispersion(\pi^\delta) - |\pi^\delta|. \bullet$$

$$(5.10)$$

We have subtracted $|\pi^\delta|$ from the sum of intra-class association and inter-class dispersion to remove the bias of goodness value of a partition. A better partition is obtained at a higher goodness value. In Example 5.4, we calculate the goodness value of each partition obtained in Example 5.3.

Example 5.4 For the first partition, we get *intra-class association* $(\pi^{0.80}) = 0.80$, *inter-class dispersion* $(\pi^{0.80}) = 0.22$, and *goodness* $(\pi^{0.80}) = 1.02$. For the second partition, we get *intra-class association* $(\pi^{0.78}) = 1.58$, *inter-class dispersion* $(\pi^{0.78}) = 0.0$, and *goodness* $(\pi^{0.78}) = 1.58$. The goodness value of the second partition is more than that of the first partition. Thus, the best non-trivial partition of $FI(1, n, \alpha)$ is $\{\{a, b\}, \{c, d\}, \{g\}\}$, and it exists at level 0.78. •

Let us look back at the databases given in Example 5.1. Items a and b appear together in most of the transactions, whenever one of them is present in a transaction. Also, items c and d appear together in most of the transactions, whenever item c or d is present in a transaction. Thus, we find that partition $\pi^{0.78}$ matches the ground reality better than partition $\pi^{0.80}$, and the output of clustering is consistent with the input. It validates the clustering technique proposed in this chapter. In the following lemma, we provide a set of necessary and sufficient conditions for the existence of a non-trivial partition.

Lemma 5.1 *Let there be n different databases. There exits a non-trivial partition of FI(1, n, α) if and only if there exists an itemset X ∈ SHFIS(1, n, α, γ) such that (1) |X| ≥ 2, and (2) SA(Y, D) ≠ SA(Z, D), for all Y, Z ∈ SHFIS(1, n, α, γ), and |Y|, |Z| ≥ 2.*

Proof We sort *SHFISs* in non-increasing order on synthesized association. Let $SA(M, D) = maximum\ \{SA(X, D): X \in HFIS(1, n, \alpha, \gamma),$ and $|X| \geq 2\}$. Before the itemset M, there exists no itemset in *SHFIS*, since it has the maximum synthesized association. Thus, it is trivially mutually exclusive with the previous *SHFIS*s. Due to condition (2), there exists a partition at the level $SA(M, D)$. In addition, the partition is non-trivial due to condition (1). This non-trivial partition contains a single class M having size ≥ 2. The remaining classes of this partition are singleton. •

At two different levels of association δ_1, and δ_2 ($\neq \delta_1$), one may obtain the same partition.

Definition 5.12 Let $C \subseteq FI(1, n, \alpha)$ be a class such that $C \neq \phi$. Two partitions π^{δ_1} and π^{δ_2} are the same if the following statement is true: $C \in \pi^{\delta_1}$ if and only if $C \in \pi^{\delta_2}$, for $\delta_1 \neq \delta_2$. •

There are $2^{|S|}$ elements in the power set of S. Also, there are two trivial partitions for a non-null set viz., (1) The partition containing singleton elements, and (2) The partition containing only one set of all elements. Thus, the number of distinct non-trivial partitions of a non-null set is always less than or equal to $2^{|S|} - 2$, for any non-null set S. In Lemma 5.2, we find the upper bound of the number of non-trivial partitions.

Lemma 5.2 *Let there be n different databases. Then the number of distinct non-trivial partitions is less than or equal to* $|\{X: |X| \geq 2 \text{ and } X \in SHFIS(D)\}|$. *Equality holds if and only if the following conditions are true.*

1. There does not exist a $X \in SHFIS(D)$, for $|X| \geq 3$.
2. $Y \cap Z = \phi$, for all $Y, Z \in SHFIS(D)$, and $|Y|, |Z| \geq 1$.
3. $SA(Y, D) \neq SA(Z, D)$, for all $Y, Z \in SHFIS(D)$, and $|Y|, |Z| \geq 2$.

Proof We arrange *SHFIS*s in a non-increasing order based on the synthesized association, for all *SHFIS*s of size greater than one. Let the arranged *SHFIS*s be X_1, $X_2, \ldots, X_m$, $m \geq 2$. There exists a partition at $SA(X_1)$, if conditions (2) and (3) are satisfied when $Y = X_1$ [Lemma 5.1]. In general, there exists another partition at $SA(X_k, D)$, if conditions (2) and (3) are satisfied for $X_1, X_2, \ldots, X_{k-1}$. If X is a *SHFIS* then $Y \subset X$ is also a *SHFIS*, for $Y \neq \phi$. Two partitions could not exist at $SA(X, D)$ and $SA(Y, D)$, since $Y \subset X$. Thus, condition (1) is necessary at the equality. The lemma follows. •

Corollary 5.1 *Let there be n different databases. The set of all non-trivial partitions of $FI(1, n, \alpha)$ is* $\{\pi^{SA(X, D)}: X \in SHFIS(D), |X| \geq 2, \text{ and } \pi^{SA(X, D)} \text{ exists}\}$. •

Based on Observation 5.1 and Corollary 5.1, one could find the level of difference in clustering levels stated as follows:

Observation 5.3 Let π be clustering of $FI(1,n, \alpha)$ using the proposed technique. Let $X = \{x_1, x_2, \ldots, x_m\} \in SHFIS (D)$ such that $\delta = SA(X, D)$, for $|X| \geq 2$. The amount of difference in clustering levels between the existing technique and the proposed technique is equal to

$$\delta - minimum\{sim_1(x_i, x_j, D) : 1 \leq i < j \leq m\}. \tag{5.11}$$

Using Algorithm 5.1, we obtain synthesized association among items of each *SHEIS* of size greater than one. We use this information for finding the best non-trivial partition of local frequency items in multiple databases (Adhikari 2013).

Algorithm 5.2 Finding the best non-trivial partition (if it exists) of local frequency items in multiple databases.
procedure *BestPartition* (*m, S*)
Input:
m: number of *SHFIS*s of size greater than one
S: array of *SHFIS*s of size greater than one
Output:
The best non-trivial partition (if it exists) of local frequency items in multiple databases
01: arrange the elements of *S* in non-increasing order on synthesized association;
02: **let** $S(m +1) = \phi$; **let** $SA(S(m +1), D) = 0$;
03: **if** ($m = 1$) **then**
04:      form a class using items in $S(1)$;
05:      **for each** item in $FI(1, n, \alpha) - S(1)$ **do**
06:          form a singleton class;
07:      **end for**
08:      a partition is formed at level $SA(S(1), D)$;
09:      return the partition;
10: **end if**
11: **let** $temp = \phi$; **let** $mutualExcl$ = false;
12: **for** $i = 1$ to m **do**
13:      **if** ($temp \cap S(i) \neq \phi$) **then** $mutualExcl$ = true; **end if**
14:      **if** (($mutualExcl$ = false) **and** ($SA(S(i), D) \neq SA(S(i +1), D)$)) **then**
15:          **for** $j = 1$ to i **do**
16:              construct a class using items in $S(j)$;
17:              $temp = temp \cup S(j)$;
18:          **end for**
19:          **for each** item in ($FI(1, n, \alpha) - temp$) **do**
20:              construct a singleton class;
21:          **end for**
22:          store the classes formed and the level of partition as $SA(S(i), D)$;
23:      **else if** ($mutualExcl$ = true) **then** go to line 26; **end if**
24:      **end if**
25: **end for**
26: return the partition having the maximum goodness;
end procedure

If *m* is equal to 1 then we have only one non-trivial partition. All the items of *SHFIS* form a class and each of the remaining frequent items forms a singleton class. The partition is formed at the level of synthesized association among items in the *SHFIS*. The variable *temp* accumulates all the items in previous *SHFIS*s. The variable *mutualExcl* is used to check the mutually exclusiveness among the current *SHFIS* and all the previous *SHFIS*s. Also, we need to check another condition expressing whether the synthesized association of current *SHFIS* different from the synthesized association among items of the next *SHFIS*. The conditions for existence of a partition are checked at line 14. If a partition exists at the current level then the items in each of the previous *SHFIS*s form a class. Each of the remaining items forms a singleton class. If the current *SHFIS* is not mutually exclusive with each of the previous *SHFIS*s of *S* then no more partitions exist. Some useful explanations are also given in Lemma 5.3.

Line 1 requires $O(m \times log(m))$ time. The *for* loop at line 5 takes $O(|FI|)$ time. The *for* loop at line 12 repeats *m* times. Let the average size of a class be *p*. The *for* loop at line 15 takes $O(m \times p)$ time. The for-loop at line 19 takes $O(|FI|)$ time. Each of the statements at line 13 and 17 takes $O(p \times |FI|)$ time for a single execution of line 11. Thus, the time complexity of the program segment lines 15–18 is $O(m \times p \times |FI|)$, for each iteration of for-loop at line 12. Also, the *for* loop at 19 takes $O(|FI|)$ time. Thus, the time complexity of program segment 12–25 is $O(m^2 \times p \times |FI|)$. Therefore, the time complexity of algorithm *Best-Partition* is $O(m^2 \times p \times |FI|)$.

Lemma 5.3 *Algorithm BestPartition finds the best partition, if such a partition exists.*

Proof We arrange *SHFIS*s of size greater than one in non-increasing order on synthesized association. Existence of only one *SHFIS* of size greater than one implies the existence of only one non-trivial partition [Lemma 5.1]. By default, it will be the best non-trivial partition.

Let there be *m* $(m \geq 2)$ *SHFIS*s of size greater than one. Let the arranged *SHFIS*s be $X_1, X_2, ..., X_m$. Then we need to check for the existence of partitions only at levels $SA(X_i, D)$, $i = 1, 2, ..., m$ [Corollary 5.1]. So, we have used a *for* loop at line 12 to check for partitions at *m* discrete levels $SA(X_i, D)$, $i = 1, 2, ...,$ *m*. At the *j*th iteration of *for* loop at line 12, we check the mutually exclusiveness of the itemset X_j with previous itemset X_k, for $k = 1, 2, ..., j - 1$. If each of $X_1, X_2,$ ..., and X_{j-1} is mutually exclusive with X_j and $SA(X_j, D) \neq SA(X_{j+1}, D)$ then the current partition is recorded. On the other hand, if each of $X_1, X_2, ...,$ and X_{j-1} is mutually exclusive with X_j and $SA(X_j, D) = SA(X_{j+1}, D)$ then the current partition is not recorded. At this point, we are not sure whether X_{j+1} is mutually exclusive with X_k, for $k = 1, 2, ..., j$. At the next iteration $i = j + 1$, the partition is recorded (if it exists), and it contains the itemsets $X_1, X_2, ..., X_{j+1}$ as classes of size greater than one and each of the remaining frequent items forms a singleton class, provided $SA(X_{j+1}, D) > SA(X_{j+2}, D)$. At the *j*-th iteration of *for* loop at line 12, if each of $X_1, X_2, ...,$ and X_{j-1} is not mutually exclusive with X_j then no more partition exists. Thus, the algorithm works correctly. •

5.5.2 Error Analysis

To evaluate the proposed clustering technique we have measured the error occurred in an experiment. Let the number of *SHFIS*s be *m*. Supports of all *SHFIS*s have been synthesized during the clustering process. There are several ways one could define error of an experiment. We have adopted the following two types of errors.

1. *Average Error*(AE) : $AE(D, \alpha, \gamma) = \dfrac{1}{m} \sum_{i=1}^{m} |SS(X_i, D) - S(X_i, D)|$ (5.12)

2. *Maximum Error* (ME) : $ME(D, \alpha, \gamma)$
$$= maximum\{|SS(X_i, D) - S(X_i, D)|, i = 1, 2, \ldots, m\}$$
(5.13)

$SS(X, D)$ and $S(X, D)$ denote synthesized support and (apriori) support of X in D respectively. Error of the experiment is relative to the number of transactions, number of items, and the length of a transaction in local databases. Thus, the error of the experiment needs to be expressed along with the average number of transactions (*ANT*), average number of items (*ANI*), and the average length of a transaction (*ALT*) in all branch databases.

5.6 Experimental Results

We have carried out many experiments to study the effectiveness of our approach. All the experiments have been implemented on a 2.8 GHz Pentium D dual processor with 512 MB of memory using Visual C++ (version 6.0) software. We present the experimental results using three real and two synthetic databases. The database *retail* (Frequent itemset mining dataset repository) is obtained from an anonymous Belgian retail supermarket store. The database *mushroom* is also available in (Frequent itemset mining dataset repository). The database *ecoli* is a subset of *ecoli database* (UCI ML repository content summary) and has been processed for the purpose of conducting experiment. Also, we have omitted non-numeric fields of *ecoli database* for the purpose of conducting experiments. Databases *random500* and *random1000* are generated synthetically for the purpose of conducting experiments. Let *NT*, *ALT*, *AFI*, and *NI* denote the number of transactions, average length of a transaction, average frequency of an item and number of items in the database, respectively. We present the characteristics of these databases in Table 5.2.

Table 5.2 Dataset characteristics

Database (*D*)	NT	ALT	AFI	NI
Retail (*R*)	88,162	11.306	99.674	10,000
Mushroom (*M*)	8,124	24.000	1,624.800	120
Ecoli (*E*)	336	7.000	25.565	92
random500 (*R5*)	10,000	6.470	109.400	500
random1000 (*R1*)	10,000	12.486	111.858	1,000

Table 5.3 Characteristics of input databases generated from R, M and E

D	NT	ALT	AFI	NI	D	NT	ALT	AFI	NI
R_0	9,000	11.244	12.070	8,384	R_5	9,000	10.856	16.710	5,847
R_1	9,000	11.209	12.265	8,225	R_6	9,000	11.200	17.416	5,788
R_2	9,000	11.337	14.597	6,990	R_7	9,000	11.155	17.346	5,788
R_3	9,000	11.490	16.663	6,206	R_8	9,000	11.997	18.690	5,777
R_4	9,000	10.957	16.040	6,148	R_9	7,162	11.692	15.348	5,456
M_0	812	24.000	295.273	66	M_5	812	24.000	221.455	88
M_1	812	24.000	286.588	68	M_6	812	24.000	216.533	90
M_2	812	24.000	249.846	78	M_7	812	24.000	191.059	102
M_3	812	24.000	282.435	69	M_8	812	24.000	229.271	85
M_4	812	24.000	259.840	75	M_9	816	24.000	227.721	86
E_0	33	7.000	4.620	50	E_5	33	7.000	3.915	59
E_1	33	7.000	5.133	45	E_6	33	7.000	3.500	66
E_2	33	7.000	5.500	42	E_7	33	7.000	3.915	59
E_3	33	7.000	4.813	48	E_8	33	7.000	3.397	68
E_4	33	7.000	3.397	68	E_9	39	7.000	4.550	60

Table 5.4 Characteristics of input databases generated from $R5$ and $R1$

D	NT	ALT	AFI	NI	D	NT	ALT	AFI	NI
$R5_0$	1,000	6.367	10.734	500	$R5_5$	1,000	6.338	10.676	500
$R5_1$	1,000	6.502	11.004	500	$R5_6$	1,000	6.624	11.248	500
$R5_2$	1,000	6.402	10.804	500	$R5_7$	1,000	6.415	10.830	500
$R5_3$	1,000	6.523	11.046	500	$R5_8$	1,000	6.579	11.158	500
$R5_4$	1,000	6.298	10.596	500	$R5_9$	1,000	6.652	11.304	500
$R1_0$	1,000	6.421	5.437	997	$R1_5$	1,000	6.444	5.454	998
$R1_1$	1,000	6.414	5.436	996	$R1_6$	1,000	6.477	5.493	997
$R1_2$	1,000	6.556	5.578	996	$R1_7$	1,000	6.477	5.487	998
$R1_3$	1,000	6.529	5.534	999	$R1_8$	1,000	6.538	5.554	997
$R1_4$	1,000	6.500	5.544	992	$R1_9$	1,000	6.500	5.561	989

Each of the above databases is divided into 10 databases for the purpose of conducting experiments. The databases obtained from R, M, E, $R5$ and $R1$ are named as R_i, M_i, E_i, $R5_i$, and $R1_i$ respectively, $i = 0, 1, \ldots, 9$. The databases R_i, M_i, E_i, $R5_i$, and $R1_i$ are called input databases, $i = 0, 1, \ldots, 9$. Some characteristics of these input databases are presented in Tables 5.3 and 5.4. We present below the results of the experiments based on the above database. We observe that some local frequency items might not get reported for some combination of input databases, α, and γ.

5.6.1 Overall Output

Using Algorithm 5.2, we have shown below the best non-trivial partition of local frequency items for each of the given databases.

(a) *Experiment with retail*: The set of frequent items in different databases are given as follows: $FI(0, 9, 0.1) = \{\{0\}, \{1\}, \{2\}, \{3\}, \{4\}, \{5\}, \{6\}, \{7\}, \{8\},$ $\{9\}, \{32\}, \{38\}, \{39\}, \{41\}, \{48\}\}$. It gives high frequency items in local databases generated from *retail* support level 0.1. *SHFIS*s of size greater than one along with their synthesized associations are given as follows: $\{39, 48\}$ (0.444), $\{39, 41, 48\}$ (0.394), $\{39, 41\}$ (0.264), $\{41, 48\}$ (0.251), $\{38, 39\}$ (0.181). The best non-trivial partition is given as follows: $\pi^{0.444} = \{\{0\}, \{1\},$ $\{2\}, \{3\}, \{4\}, \{5\}, \{6\}, \{7\}, \{8\}, \{9\}, \{32\}, \{38\}, \{41\}, \{39, 48\}\}$. We observe that items 39 and 48 are highly associated, and their estimated association is 0.444. Also, there exists high association among items 39, 41 and 48.

(b) *Experiment with mushroom*: The set of frequent items in different databases are given as follows: $FI(0, 9, 0.5) = \{\{1\}, \{2\}, \{3\}, \{6\}, \{7\}, \{9\}, \{10\}, \{11\},$ $\{23\}, \{24\}, \{28\}, \{29\}, \{34\}, \{36\}, \{37\}, \{38\}, \{39\}, \{48\}, \{52\}, \{53\}, \{54\},$ $\{56\}, \{58\}, \{59\}, \{61\}, \{63\}, \{66\}, \{67\}, \{76\}, \{85\}, \{86\}, \{90\}, \{93\}, \{94\},$ $\{95\}, \{99\}, \{101\}, \{102\}, \{110\}, \{114\}, \{116\}, \{117\}, \{119\}\}$. We have taken α quite high, since the database is dense. Items are repeated in many transactions. As a result, associations among some items are very high. Top ten *SHFIS*s of size greater than one along with their synthesized associations are given as follows: $\{34, 90\}$ (0.9999), $\{34, 86\}$ (0.9995), $\{34, 85\}$ (0.9956), $\{34, 36, 85\}$ (0.9897), $\{34, 36, 90\}$ (0.9879), $\{34, 85, 90\}$ (0.9801), $\{34, 36, 86\}$ (0.9778), $\{34, 86, 90\}$ (0.9774), $\{34, 85, 86\}$ (0.9683), $\{85, 86\}$ (0.9627). The transactions in different databases are highly similar, since two transactions in a database have many common items. Therefore, the best partition has been reported at very high level of association. The best non-trivial partition is given as follows: $\pi^{0.9999} = \{\{1\}, \{2\}, \{3\}, \{6\}, \{7\}, \{9\}, \{10\}, \{11\}, \{23\},$ $\{24\}, \{28\}, \{29\}, \{36\}, \{37\}, \{38\}, \{39\}, \{48\}, \{52\}, \{53\}, \{54\}, \{56\}, \{58\},$ $\{59\}, \{61\}, \{63\}, \{66\}, \{67\}, \{76\}, \{85\}, \{86\}, \{93\}, \{94\}, \{95\}, \{99\},$ $\{101\}, \{102\}, \{110\}, \{114\}, \{116\}, \{117\}, \{119\}, \{34, 90\}\}$.

(c) *Experiment with ecoli*: Dataset *ecoli* has fewer items. Also, the number of transaction is less. Therefore, the supports of different itemsets are relatively higher. The frequent items in different databases are given as follows: $FI(0, 9,$ $0.1) = \{\{0\}, \{20\}, \{23\}, \{24\}, \{25\}, \{26\}, \{27\}, \{28\}, \{29\}, \{30\}, \{31\},$ $\{32\}, \{33\}, \{34\}, \{35\}, \{36\}, \{37\}, \{38\}, \{39\}, \{40\}, \{41\}, \{42\}, \{43\}, \{44\},$ $\{45\}, \{46\}, \{47\}, \{48\}, \{49\}, \{50\}, \{51\}, \{52\}, \{54\}, \{56\}, \{57\}, \{58\}, \{59\},$ $\{61\}, \{63\}, \{64\}, \{65\}, \{66\}, \{67\}, \{68\}, \{69\}, \{70\}, \{71\}, \{72\}, \{73\}, \{74\},$ $\{75\}, \{76\}, \{77\}, \{78\}, \{79\}, \{80\}, \{81\}, \{92\}, \{100\}\}$. Some itemsets such as $\{37\}, \{48\}, \{50\}, \{52\}$, and $\{48, 50\}$ has very high support. Top ten *SHFIS*s of size greater than one along with their synthesized associations are given as follows: $\{48, 50\}$(0.8036), $\{37, 48, 50\}$(0.2321), $\{48, 50, 52\}$ (0.2292), $\{40, 48, 50\}$(0.2262), $\{44, 48, 50\}$(0.2262), $\{46, 48, 50\}$(0.1935),

{37, 48}(0.1905), {44, 50}(0.1905), {48, 50, 51}(0.1905), {40, 48}(0.1845). The best non-trivial partition is given as follows: $\pi^{0.8036} = \{\{0\}, \{20\}, \{23\},$ {24}, {25}, {26}, {27}, {28}, {29}, {30}, {31}, {32}, {33}, {34}, {35}, {36}, {37}, {38}, {39}, {40}, {41}, {42}, {43}, {44}, {45}, {46}, {47}, {49}, {51}, {52}, {54}, {56}, {57}, {58}, {59}, {61}, {63}, {64}, {65}, {66}, {67}, {68}, {69}, {70}, {71}, {72}, {73}, {74}, {75}, {76}, {77}, {78}, {79}, {80}, {81}, {92}, {100}, {48, 50}}. The best partition has occurred at very high level.

(d) *Experiment with random500*: Items in *random500* are sparsely distributed. Therefore, a very few items are reported even at support level 0.02. The set of frequent items in different databases are given as follows: *FI*(0, 9, 0.02) = {{3}, {4}, {8}, {25}, {32}, {101}, {145}, {178}, {221}, {234}, {256}, {289}, {320}, {442}}. *SHFISs* of size greater than one along with their synthesized associations are given as follows: {3, 25} (0.073), {3, 4} (0.068), {8, 25} (0.057), {25, 32} (0.048), {101, 178} (0.035). The best non-trivial partition is given as follows: $\pi^{0.073} = \{\{4\}, \{8\}, \{32\}, \{101\}, \{145\}, \{178\}, \{221\},$ {234}, {256}, {289}, {320}, {442}, {3, 25}}. The best partition occurs at a low level of 0.073.

(e) *Experiment with random1000*: Dataset *random1000* is also sparsely distributed. So, very few items are reported at support level 0.01. The set of frequent items in different databases are given as follows: *FI*(0, 9, 0.01) = {{1}, {6}, {9}, {20}, {25}, {46}, {79}, {95}, {122}, {136}, {180}, {234}, {268}, {291}, {325}, {378}, {406}, {432}}. *SHFISs* of size greater than one along with their synthesized associations are given as follows: {1, 20} (0.052), {1, 9} (0.034), {25, 46} (0.030), {46, 95} (0.027), {122, 180} (0.021), {180, 234} (0.016), {291, 378} (0.015). The best non-trivial partition is given as follows: $\pi^{0.052} = $ {{6}, {9}, {25}, {46}, {79}, {95}, {122}, {136}, {180}, {234}, {268}, {291}, {325}, {378}, {406}, {432}, {1, 20}}. The best partition occurs at a low level of 0.052.

5.6.2 Synthesis of High Frequency Itemsets

Partitions are based on high frequency itemsets in multiple databases. An itemset may not get reported from all the n databases. Thus, it is required to estimate its support in the union of n databases. In the Tables 5.5, 5.6, 5.7, 5.8 and 5.9, we have presented a few itemsets and their errors of synthesizing supports in the union on all databases.

(a) *Experiment with retail*: In Table 5.5, we present errors in synthesizing *HFISs*. The dataset *retail* has a unique characteristic that some itemsets such as {32}, {38}, {39}, {48}, {38, 39} and {39, 48} are extracted from every branch database. Thus, the error in synthesizing support of each of the above *HFISs* is zero. Their supports are exact in the union of all the branch databases.

Table 5.5 Error in synthesizing supports of *HFIS*s in *retail* at $\alpha = 0.1$ and $\gamma = 0.6$ (least 15)

HFIS X	|SS(X, R) − S(X, R)|	HFIS X	|SS(X, R) − S(X, R)|	HFIS X	|SS(X, R) − S(X, R)|
{32}	0.0000	{39, 48}	0.0000	{3}	0.0020
{39}	0.0000	{9}	0.0017	{4}	0.0020
{38}	0.0000	{5}	0.0019	{7}	0.0020
{48}	0.0000	{6}	0.0019	{8}	0.0020
{38, 39}	0.0000	{2}	0.0019	{1}	0.0021

Table 5.6 Error in synthesizing supports of *HFIS*s in *mushroom* at $\alpha = 0.5$ and $\gamma = 0.6$ (least 15)

HFIS X	|SS(X, M) − S(X, M)|	HFIS X	|SS(X, M) − S(X, M)|	HFIS X	|SS(X, M) − S(X, M)|
{67}	0.0010	{53, 85}	0.0011	{86, 90}	0.0013
{34, 67}	0.0010	{53, 90}	0.0011	{39}	0.0018
{53, 86}	0.0010	{53}	0.0012	{24, 90}	0.0022
{34, 39}	0.0011	{85, 86}	0.0013	{24, 85}	0.0033
{34, 90}	0.0011	{85, 90}	0.0013	{24, 86}	0.0033

Table 5.7 Error in synthesizing supports of *HFIS*s in *ecoli* at $\alpha = 0.1$ and $\gamma = 0.6$ (least 15)

HFIS X	|SS(X, E) − S(X, E)|	HFIS X	|SS(X, E) − S(X, E)|	HFIS X	|SS(X, E) − S(X, E)|
{40, 50}	0.0013	{40}	0.0018	{50, 51}	0.0024
{50, 52}	0.0013	{48, 50, 52}	0.0018	{37, 48, 50}	0.0030
{37}	0.0014	{44, 48, 50}	0.0021	{40, 48, 50}	0.0033
{52}	0.0014	{44}	0.0024	{46, 48, 50}	0.0035
{48, 51}	0.0015	{46, 48}	0.0024	{48, 50, 51}	0.0037

Table 5.8 Error in synthesizing supports of *HFIS*s in *random500* at $\alpha = 0.02$ and $\gamma = 0.6$ (least 15)

HFIS X	|SS(X, R) − S(X, R)|	HFIS X	|SS(X, R) − S(X, R)|	HFIS X	|SS(X, R) − S(X, R)|
{4}	0.0009	{178}	0.0023	{3, 4}	0.0026
{3}	0.0011	{221}	0.0025	{256}	0.0028
{8}	0.0014	{234}	0.0025	{3, 25}	0.0028
{25}	0.0020	{101}	0.0026	{25, 32}	0.0031
{3, 25}	0.0023	{8, 25}	0.0026	{101, 178}	0.0035

(b) *Experiment with mushroom*: The transactions in local databases are highly similar, in the sense that two transactions in a database have many common items. Thus, we get many frequent itemsets in local databases even at a high

Table 5.9 Error in synthesizing supports of *HFIS*s at *random1000* at $\alpha = 0.02$ and $\gamma = 0.6$ (least 15)

HFIS X	|SS(X, R) − S(X, R)|	HFIS X	|SS(X, R) − S(X, R)|	HFIS X	|SS(X, R) − S(X, R)|
{9}	0.0000	{291}	0.0018	{25, 46}	0.0028
{1}	0.0008	{268}	0.0021	{46, 95}	0.0028
{79}	0.0012	{136}	0.0023	{122, 180}	0.0030
{406}	0.0015	{46}	0.0026	{291, 378}	0.0033
{122}	0.0018	{1, 9}	0.0028	{180, 234}	0.0033

value of α. The errors in synthesizing some *HFIS*s are presented in Table 5.6. No itemset has been reported from all the local databases. But, there exist many itemsets that are extracted from eight/nine out of ten local databases.

(c) *Experiment with ecoli*: The average size of databases and the average length of transactions in different databases from *ecoli* are smaller. Thus, the error of synthesizing a *HFIS* is relatively higher. The errors in synthesizing some *HFIS*s are presented in Table 5.7. No itemset has been reported from all the local databases. But, there exist many itemsets that are extracted from eight/nine out of ten local databases.

(d) *Experiment with random500*: No itemset has been reported from all the local databases. The databases generated from *random500* are uniform. Many synthesized itemsets extracted have similar supports in multiple databases. In Table 5.8, we present errors in synthesizing *HFIS*s in *random500*.

(e) *Experiment with random1000*: The databases generated from *random1000* are uniform. Many synthesized itemsets extracted have similar supports in multiple databases. We observe that itemset {9} is extracted from every branch database. Thus, the error in synthesizing support of {9} is zero. In Table 5.9, we present errors in synthesizing *HFIS*s in *random1000*.

5.6.3 Error Quantification

The error obtained in the experiment is based on the definitions given in (5.12) and (5.13). The errors reported in different experiments are presented in Table 5.10.

Table 5.10 Error of the experiments

D	α	γ	AE (ANT, ALT, ANI)	ME (ANT, ALT, ANI)
$\bigcup_{i=0}^{9} R_i$	0.1	0.7	0.0012 (8,816.2, 11.31, 5,882.1)	0.0029 (8816.2, 11.31, 5882.1)
$\bigcup_{i=0}^{9} M_i$	0.5	0.7	0.0013 (812.4, 24.00, 80.7)	0.0038 (812.4, 24.00, 80.7)
$\bigcup_{i=0}^{9} E_i$	0.5	0.7	0.0013 (33.6, 7.00, 56.5)	0.0041 (33.6, 7.00, 56.5)
$\bigcup_{i=0}^{9} R5_i$	0.02	0.7	0.0010 (1,000, 6.47, 500)	0.0042 (1,000, 6.47, 500)
$\bigcup_{i=0}^{9} R1_i$	0.01	0.7	0.0009 (1,000, 6.48, 995.9)	0.0038 (1,000, 6.48, 995.9)

If the average number of transactions in different databases increases then the average error of synthesizing *HEIS*s is likely to decrease, provided the average length of transactions and the average number of items remain constant. If the average length of transactions in different databases increases then the average error of synthesizing *HEIS*s is likely to increase, provided the average number of transactions and the average number of items remain constant. Lastly, if the average number of items in different databases increases then the error of synthesizing *HEIS*s is likely to decrease, provided the average number of transactions and the average length of transactions remain constant.

5.6.4 Average Error Versus γ

Here γ represents threshold of minimum number of extractions from different data sources. We have conducted experiments to study the behaviour of AE over different γs. The purpose of the experiment would be lost if we keep γ at a high value, since the number of *HEIS*s also decreases as γ increases. Thus, a decision based on *HEIS*s would have low validity at a high value of γ. In Figs. 5.1, 5.2, 5.3, 5.4 and 5.5, we present graphs of AE plotted versus γ obtained in different experiments.

For databases generated from *retail*, the same itemsets are reported from all the local databases when γ is greater than or equal to 0.9. As a result, the AE becomes 0. In that case, there is no need to estimate the support of an itemset.

For datasets *retail* and *mushroom*, the values of AE decrease gradually when γ ∈ [0.4, 0.7]. But, AE falls at a faster rate when γ ∈ [0.8, 0.9].

Fig. 5.1 AE versus γ at α = 0.1 (*retail*)

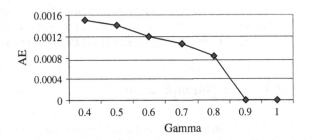

Fig. 5.2 AE versus γ at α = 0.5 (*mushroom*)

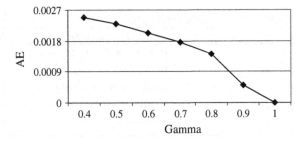

Fig. 5.3 AE versus γ at
$\alpha = 0.1$ (*ecoli*)

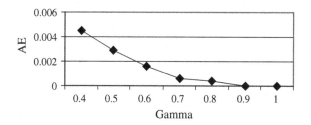

Fig. 5.4 AE versus γ at
$\alpha = 0.02$ (*random500*)

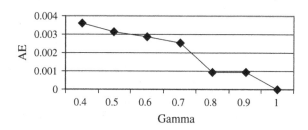

Fig. 5.5 AE versus γ at
$\alpha = 0.01$ (*random1000*)

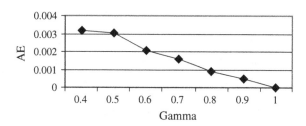

Also, for databases generated from *ecoli* dataset, the same itemsets are reported from all the local databases when γ is greater than or equal to 0.9. AE falls gradually when γ varies from 0.4 to 0.9.

The behaviours of graph AE versus γ remain more or less the same for databases generated from *random500 random1000*. In general, we find that AE decreases as γ increases. In an extreme case, when γ is 1.0, no itemset has become frequent in all the local databases. Thus, AE is becomes 0 by default. From the figures presented above, we find that the value of γ around 0.7 would have been a good choice for clustering frequent items in different databases.

5.6.5 Average Error Versus α

Here α is user-defined level of minimum support. We have also conducted experiments to study the behaviour of AE over different αs. In Figs. 5.6, 5.7, 5.8, 5.9 and 5.10, we present graphs of AE versus α for different experiments.

Fig. 5.6 AE versus α at
γ = 0.7 (*retail*)

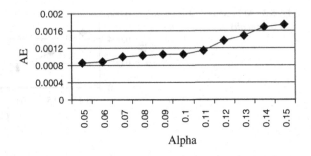

Fig. 5.7 AE versus α at
γ = 0.7 (*mushroom*)

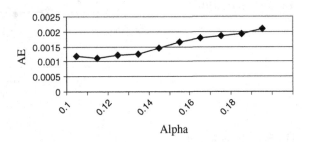

Fig. 5.8 AE versus α at
γ = 0.7 (*ecoli*)

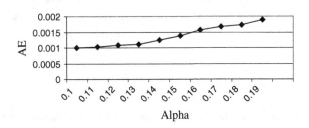

Fig. 5.9 AE versus α at
γ = 0.7 (*random500*)

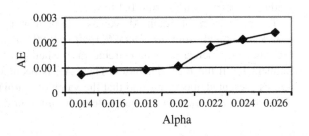

Fig. 5.10 AE versus α at
γ = 0.7 (*random1000*)

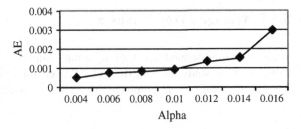

In each of these experiments, we have taken around ten values of α. In dataset *retail*, itemsets are reported at a moderate value of α. Here we have started α at 0.05.

As noted earlier, the dataset *mushroom* is dense. Itemsets are reported even at a higher value of α. Here we have started α at 0.1.

Dataset *ecoli* is small. Thus, itemsets are reported at a higher value of α. Here we have started α at 0.1. In all the above graphs, we observe a slow and steady increase of the values of AE when α increases.

The behaviour of graph AE versus α remains more or less the same for databases generated from *random500* and *random1000*, since all these datasets are synthetic. As α increases, the larger number of databases would fail to extract an itemset. Thus, the error of synthesizing an itemset is likely to increase with the increase of α. In general, we find that AE of the experiment gets higher as α increases.

5.6.6 Clustering Time Versus Number of Databases

We have also studied the behaviour of clustering time required over the number of databases used in an experiment. In Figs. 5.11, 5.12, 5.13, 5.14 and 5.15, we present graphs of clustering time versus number of databases used in different experiments.

Although the number of transactions is less for *mushroom* as compared to *retail*, but the number of items in a transaction in *mushroom* is almost double than that of *retail*. As a result, graphs in Figs. 5.11 and 5.12 become comparable.

The sizes of local databases generated from *ecoli* are the smallest. Thus, the increment of clustering time is the least among all the five experiments. There is an increment of 0.17 s, on an average, for processing an extra database generated from *ecoli*.

Fig. 5.11 Clustering time versus number of databases at $\alpha = 0.1$ and $\gamma = 0.7$ (*retail*)

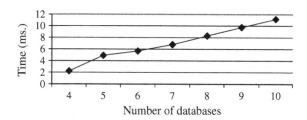

Fig. 5.12 Clustering time versus number of databases at $\alpha = 0.5$ and $\gamma = 0.7$ (*mushroom*)

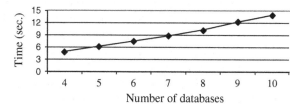

Fig. 5.13 Clustering time versus number of databases at $\alpha = 0.5$ and $\gamma = 0.7$ (*ecoli*)

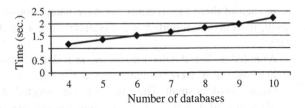

Fig. 5.14 Clustering time versus number of databases at $\alpha = 0.02$ and $\gamma = 0.7$ (*random500*)

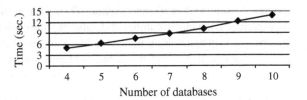

Fig. 5.15 Clustering time versus number of databases at $\alpha = 0.01$ and $\gamma = 0.7$ (*random1000*)

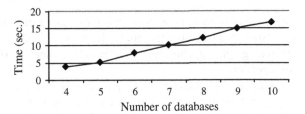

The graph of clustering time using dataset *random1000* is steeper than that of dataset *random500*, since the local databases generated from *random1000* are larger. As the number of databases increases, the number of frequent itemsets also increases. In general, we find that clustering time increases as the number of databases increases.

5.6.7 Comparison with Existing Technique

We have explained in Example 2 why the existing technique (Wu et al. 2005; Adhikari and Rao 2008c) could not cluster the local frequency items with high accuracy. Also, we derived an expression for difference occurring at clustering levels, see Observation 5.3. We observe that the proposed clustering technique is likely to cluster local frequency items at higher level, if a high frequency itemset of size greater than two has been synthesized. In each of Tables 5.11, 5.12, 5.13, 5.14 and 5.15, we present and compare partitions using the proposed technique and the existing technique, where δ is the level of clustering.

In dataset *retail*, partitions are reported when $\alpha = 0.1$ and $\gamma = 0.7$. Best partition is judged using the goodness measure as stated in (5.13). The dataset *retail*

Table 5.11 A sample clustering of local frequency items in multiple databases (*retail*)

α	γ	Partition	δ (existing approach)	δ (proposed approach)	Difference in clustering levels
0.1	0.7	{{0}, {1}, {2}, {3}, {4}, {5}, {6}, {7}, {8}, {9}, {32}, {38, 39}, {39, 41, 48}}	0.251	0.394	0.143

Table 5.12 A sample clustering of frequent items in multiple databases (*mushroom*)

α	γ	Partition	δ (existing approach)	δ (proposed approach)	Difference in clustering levels
0.5	0.7	{{1}, {2}, {3}, {6}, {7}, {9}, {10}, {11}, {23}, {24}, {28}, {29}, {37}, {38}, {39}, {48}, {52}, {53}, {54}, {56}, {58}, {59}, {61}, {63}, {66}, {67}, {76}, {86}, {90}, {93}, {94}, {95}, {99}, {101}, {102}, {110}, {114}, {116}, {117}, {119}, {34, 90}, {34, 86}, {34, 36, 85}}	0.843	0.990	0.147

Table 5.13 A sample clustering of frequent items in multiple databases (*ecoli*)

α	γ	Partition	δ (existing approach)	δ (proposed approach)	Difference in clustering levels
0.1	0.7	{{0}, {20}, {23}, {24}, {25}, {26}, {27}, {28}, {29}, {30}, {31}, {32}, {33}, {34}, {35}, {36}, {38}, {39}, {40}, {41}, {42}, {43}, {44}, {45}, {46}, {47}, {49}, {51}, {52}, {54}, {56}, {57}, {58}, {59}, {61}, {63}, {64}, {65}, {66}, {67}, {68}, {69}, {70}, {71}, {72}, {73}, {74}, {75}, {76}, {77}, {78}, {79}, {80}, {81}, {92}, {100}, {37, 48, 50}}	0.179	0.232	0.053

Table 5.14 A sample clustering of frequent items in multiple databases (*random500*)

α	γ	Partition	δ (existing approach)	δ (proposed approach)	Difference in clustering levels
0.1	0.7	{{4}, {8}, {32}, {101}, {145}, {178}, {221}, {234}, {256}, {289}, {320}, {442}, {3, 25}}	0.073	0.073	0.0

reports synthesized 3-itemsets even at a moderate value of α. In most of the cases, the synthesized support is higher than the maximum of synthesized supports of all the three 2-itemsets generated from it.

Table 5.15 A sample clustering of frequent items in multiple databases (*random1000*)

α	γ	Partition	δ (existing approach)	δ (proposed approach)	Difference in clustering levels
0.1	0.7	{{6}, {9}, {25}, {46}, {79}, {95}, {122}, {136}, {180}, {234}, {268}, {291}, {325}, {378}, {406}, {432}, {1, 20}}	0.052	0.052	0.0

The dataset *mushroom* is dense. Many synthesized 3-itemsets are reported from multiple databases generated from *mushroom*. Thus, partitions exists at high value of $\alpha = 0.5$.

The dataset *ecoli* is small. Many synthesized 3-itemsets are reported from multiple databases generated from *ecoli*. Thus, partitions exists even at moderate value of $\alpha = 0.1$.

The datasets *random500* and *random1000* are sparse. No k-itemset is reported ($k \geq 3$) when $\alpha = 0.1$ and $\gamma = 0.7$. We observe that the existing approach (Wu et al. 2005; Adhikari and Rao 2008c) and proposed approach produce the same result when a partition contains a class of maximum size 2. As a result, both the clustering techniques cluster items at the same level.

5.7 Conclusions

Clustering relevant objects is an important task in many decision support systems. In this chapter, we have proposed a new clustering technique for clustering local frequency items in multiple databases. For the purpose of applying the new clustering technique, we have proposed an algorithm for synthesizing high frequency itemsets. We have observed that an existing clustering technique might cluster local frequency items at a low level even when the associations among items in the itemsets are high. But the proposed clustering technique might cluster local frequency items at a higher level by capturing association among items using measure A_2. Especially, when the items in a database are highly associated, the proposed clustering technique clusters items at a higher level. The main problem with an existing clustering technique is that it might not be able to estimate association among items in a class with high accuracy. Thus, it might fail to cluster local frequency items at the higher level. We have shown that the proposed clustering method is sound. In case of dense databases, there is a considerable difference in clustering levels between the partitions made by the existing method and the proposed method. This is evident from the experimental results presented in Tables 5.10, 5.11 and 5.12. To apply the proposed technique, it does not matter whether the databases are required to be generated by real applications or synthetic database generators, since the algorithm is not dependent on how the databases were formed. The experimental results show that the proposed clustering technique is effective.

References

Adhikari A (2012) Synthesizing global exceptional patterns in different data sources. J Intell Syst 21(3):293–323

Adhikari A (2013) Clustering local frequency items in multiple databases. Inf Sci 237:221–241

Adhikari A, Rao PR (2008a) Capturing association among items in a database. Data Knowl Eng 67(3):430–443

Adhikari A, Rao PR (2008b) Mining conditional patterns in a database. Pattern Recogn Lett 29(10):1515–1523

Adhikari A, Rao PR (2008c) Efficient clustering of databases induced by local patterns. Decis Support Syst 44(4):925–943

Adhikari A, Rao PR (2008d) Synthesizing heavy association rules from different real data sources. Pattern Recogn Lett 29(1):59–71

Adhikari J, Rao PR, Adhikari A (2009) Clustering items in different data sources induced by stability. Int Arab J Inf Technol 6(4):394–402

Adhikari A, Ramachandrarao P, Pedrycz W (2010a) Developing multi-database mining applications. Springer, London

Adhikari A, Rao PR, Prasad B, Adhikari J (2010b) Mining multiple large data sources. Int Arab J Inf Technol 7(2):243–251

Adhikari A, Ramachandrarao P, Pedrycz W (2011a) Study of select items in different data sources by grouping. Knowl Inf Syst 27(1):23–43

Adhikari J, Rao PR, Pedrycz W (2011b) Mining icebergs in time-stamped databases. In: Proceedings of Indian international conferences on artificial intelligence, pp 639–658

Agrawal R, Shafer J (1996) Parallel mining of association rules. IEEE Trans Knowl Data Eng 8(6):962–969

Agrawal R, Srikant R (1994) Fast algorithms for mining association rules. In: Proceedings of the international conference on very large data bases, pp 487–499

Aggarwal C, Yu P (1998) A new framework for itemset generation. In: Proceedings of PODS, pp 18–24

Agrawal R, Imielinski T, Swami A (1993) Mining association rules between sets of items in large databases. In: Proceedings of SIGMOD conference on management of data, pp 207–216

Ali K, Manganaris S, Srikant R (1997) Partial classification using association rules. In: Proceedings of the 3rd international conference on knowledge discovery and data mining, pp 115–118

Brin S, Motwani R, Ullman JD, Tsur S (1997) Dynamic itemset counting and implication rules for market basket data. In: Proceedings of SIGMOD conference, pp 255–264

Chattratichat J, Darlington J, Ghanem M, Guo Y, Hüning H, Köhler M, Sutiwaraphun J, To HW, Yang D (1997) Large scale data mining: challenges and responses. In: Proceedings of the third international conference on knowledge discovery and data mining, pp 143–146

Chen L, Zou L, Tu L (2012) A clustering algorithm for multiple data streams based on spectral component similarity. Inf Sci 183(1):35–47

Cheung D, Ng V, Fu A, Fu Y (1996) Efficient mining of association rules in distributed databases. IEEE Trans Knowl Data Eng 8(6):911–922

Duan L, Street WN (2009) Finding maximal fully-correlated itemsets in large databases. In: Proceedings of ICDM, pp 770–775

Estivill-Castro V, Yang J (2004) Fast and robust general purpose clustering algorithms. Data Min Knowl Disc 8(2):127–150

Frequent itemset mining dataset repository. http://fimi.cs.helsinki.fi/data/

Frequent itemset mining implementations repository. http://fimi.cs.helsinki.fi/src/

Han J, Kamber M (2001) Data mining: concepts and techniques. Morgan Kauffmann Publishers, San Francisco

Han J, Pei J, Yiwen Y (2000) Mining frequent patterns without candidate generation. In: Proceedings of SIGMOD conference on management of data, pp 1–12

Hershberger SL, Fisher DG (2005) Measures of association, encyclopedia of statistics in behavioral science. Wiley, London

He D, Wu X, Zhu X (2010) Rule synthesizing from multiple related databases. In: Proceedings of PAKDD(2), pp 201–213

Jain AK, Murty MN, Flynn PJ (1999) Data clustering: a review. ACM Comput Surv 31(3):264–323

Lee J-S, Ólafsson S (2011) Data clustering by minimizing disconnectivity. Inf Sci 181(4):732–746

Liu CL (1985) Elements of discrete mathematics. McGraw-Hill, New York

Liu H, Lu H, Yao J (2001) Toward multi-database mining: identifying relevant databases. IEEE Trans Knowl Data Eng 13(4):541–553

Malinen MI, Fränti P (2012) Clustering by analytic functions. Inf Sci 217:31–38

Mampaey M, Vreeken J (2013) Summarizing categorical data by clustering attributes. Data Min Knowl Disc 26(1):130–173

Piatetsky-Shapiro G (1991) Discovery, analysis, and presentation of strong rules. In: Proceedings of knowledge discovery in databases, pp 229–248

Savasere A, Omiecinski E, Navathe S (1995) An efficient algorithm for mining association rules in large databases. In: Proceedings of the 21st international conference on very large data bases, pp 432–443

Tan P-N, Kumar V, Srivastava J (2003) Selecting the right interestingness measure for association patterns. In: Proceedings of SIGKDD conference, pp 32–41

UCI ML repository content summary. http://www.ics.uci.edu/~mlearn/MLSummary.html

Wu X, Zhang S (2003) Synthesizing high-frequency rules from different data sources. IEEE Trans Knowl Data Eng 14(2):353–367

Wu X, Zhang C, Zhang S (2005) Database classification for multi-database mining. Inf Syst 30(1):71–88

Yakut I, Polat H (2012) Privacy-preserving hybrid collaborative filtering on cross distributed data. Knowl Inf Syst 30(2):405–433

Zhang C, Liu M, Nie W, Zhang S (2004a) Identifying global exceptional patterns in multi-database mining. IEEE Comput Intell Bull 3(1):19–24

Zhang T, Ramakrishnan R, Livny M (1997) BIRCH: a new data clustering algorithm and its applications. Data Min Knowl Disc 1(2):141–182

Zhang S, Wu X, Zhang C (2003) Multi-database mining. IEEE Comput Intell Bull 2(1):5–13

Zhang S, Zhang C, Wu X (2004b) Knowledge discovery in multiple databases. Springer, London

Zhou W, Xiong H (2009) Efficient discovery of confounders in large data sets. In: Proceedings of ICDM, pp 647–656

Chapter 6
Mining Patterns of Select Items
in Different Data Sources

A number of important decisions are based on a set of specific items in a database called *select items*. Thus the analysis of select items in multiple databases becomes of primordial relevance. In this chapter, we focus on the following issues. First, a model of mining global patterns of select items from multiple databases is presented. Second, a measure of quantifying an overall association between two items in a database is discussed. Third, we present an algorithm that is based on the proposed overall association between two items in a database for the purpose of grouping the frequent items in multiple databases. Each group contains a select item called the *nucleus item* and the group grows while being centered around the nucleus item. Experimental results are concerned with some synthetic and real-world databases.

6.1 Introduction

In Chap. 4, we have presented a generalized multi-database mining technique, MDMT: PFT+SPS, by combining pipelined feedback technique (PFT) and simple pattern synthesizing (SPS) algorithm. We have noted that one could develop a multi-database mining application using MDMT: PFM+SPS which performs reasonably well. The following question arises as to whether MDMT: PFM+SPS is the most suitable technique for mining multiple large databases in all situations. In many applications, one may need to extract true non-local patterns of a set of specific items present in multiple large databases. In such applications, MDMT: PFM+SPS could not be fully endorsed as it may return approximate non-local patterns. In this chapter, we present a technique that extracts genuine global patterns of a set of specific items from multiple large databases.

Many decisions are based on a set of specific items called *select items*. Let us highlight several decision support applications where the decisions are based on the performance of select items.

A. Adhikari et al., *Data Analysis and Pattern Recognition in Multiple Databases*, Intelligent Systems Reference Library 61, DOI: 10.1007/978-3-319-03410-2_6,

- Consider a set of items (products) that are profit making. We could consider them as the select items in this context. Naturally, the company would like to promote them. There are various ways one could promote an item. An indirect way of promoting a select item is to promote items that are positively associated with it. The implication of positive association between a select item P and another item Q is that if Q is purchased by a customer then P is likely to be purchased by the same customer at the same time. In this way, item P becomes indirectly promoted. It is important to identify the items that are positively associated with a select item.
- Each of the select items could be of high standard. Thus, they bring goodwill for the company. They help promoting other items. Therefore it is essential to know how the sales of select items affect other items. Before proceeding with such analyses, one may need to identify the items that are positively associated with the select items.
- Again, each of the select items could be a low-profit making product. From this perspective, it is important to know how they promote the sales of other items. Otherwise, the company could stop dealing with those products.

In general, the performance of select items could affect many decision making problems. Thus a better, more comprehensive analysis of select items might lead to better decisions. We study the select items based on the frequent itemsets extracted from multiple databases. The first question is whether a "traditional" data mining technique could provide a good solution when dealing with multiple large databases. The "traditional" way of mining multiple databases might not provide a sound solution due to several reasons:

- The company might have to employ parallel hardware and software to deal with a sheer volume of data.
- A single computer might take unreasonable amount of time to mine a large volume of data. In some extreme cases, it might not be feasible to carry data mining.
- A traditional data mining algorithm might extract a large number of patterns comprising many irrelevant items. Thus the processing of patterns could be complex and time consuming.

Therefore, the traditional way of mining multiple databases could not provide an efficient solution to the problem. In this situation, one could apply local pattern analysis (Zhang et al. 2003). Given this model of mining multiple databases, each branch of a company requires to mine its local database by utilizing some traditional data mining technique. Afterwards, each branch forwards the pattern base to the central office. The central office processes such pattern bases collected from different branches and synthesizes the global patterns and eventually makes decisions. Due to the reasons stated above, the local pattern analysis would not be a judicious choice to solve the proposed problem.

Each local pattern base might contain a large number of patterns consisting of many irrelevant items. Under these circumstances, the data analysis becomes complicated and time consuming. A pattern of a select item might be absent in some local pattern bases. One may be required to estimate or ignore some patterns in certain databases. Therefore we may fail to report the true global patterns of select items in the union of all local databases. All in all, we conclude that the local pattern analysis alone might not provide a good solution to the problem.

Due to difficulties identified above, we aim at developing a technique that mines *true* global patterns of select items in multiple databases. There are two apparent advantages of using such technique. First, the synthesized global patterns are exact. In other words, there is no necessity to estimate patterns in some databases. Second, one could avoid dealing with huge volumes of data.

6.2 Mining Global Patterns of Select Items

In Fig. 6.1, we visualize an essence of the technique of mining global patterns of select items in multiple databases (Adhikari et al. 2011). It consists of the following steps:

1. Each branch constructs the database and sends it to the central office.
2. Also, each branch extracts patterns from its local database.
3. The central office amalgamates these forwarded databases into a single database *FD*.

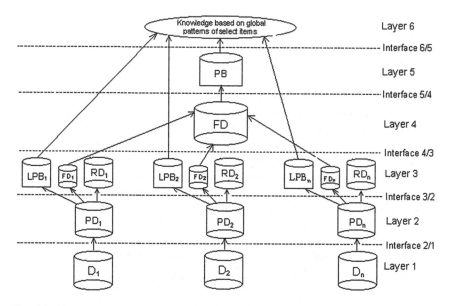

Fig. 6.1 A multilevel process of mining global patterns of select items in multiple databases

4. A traditional data mining technique is applied to extract patterns from *FD*.
5. The global patterns of select items could be extracted effectively from local patterns and the patterns extracted from *FD*.

In Sect. 6.4, we will explain steps 1–5 with the help of a specific illustrative example. The local databases are located at the bottom level of the figure. We need to process these databases as they may not be at the appropriate state to realize the mining task. Various data preparation techniques (Pyle 1999) like data cleaning, data transformation, data integration, data reduction, etc. are applied to the data present in local databases. We produce local processed database PD_i for the i-th branch, for $i = 1, 2,..., n$. The proposed model comes with a set of interfaces combined with a set of layers. Each interface delivers a set of operations that produces dataset(s) (or knowledge) based on the dataset(s) available at the lower level. There are five interfaces in the proposed model. The functions of the interfaces are described below.

Interface 2/1 is used to clean/transform/integrate/reduce data present at the lowest level. By applying these procedures we construct database resulting from the original database. These operations are carried out at the respective branch. We apply an algorithm (located at interface 3/2) to partition a local database into two parts: forwarded database and remaining database. It is easy to find the forwarded database corresponding to a given database. In the following paragraph, we discuss how to construct FD_i, from D_i, $i = 1, 2,..., n$.

Initially, FD_i is kept empty. Let T_{ij} be the j-the transaction of D_i, $j = 1, 2,..., |D_i|$. For D_i, a for-loop on j would run $|D_i|$ times. At the j-th iteration, the transaction T_{ij} is tested. If T_{ij} contains at least one of the select items then FD_i is updated, resulting in the union $FD_i \cup \{T_{ij}\}$. At the end of the for-loop completed for j, FD_i is constructed.

A transaction related to select items might contain items other than those being selected. A traditional data mining algorithm could be applied to extract patterns from *FD*. Let *PB* be the pattern base returned by a traditional data mining algorithm (at the interface 5/4). Since the database *FD* is not large, one could reduce further the values of user-defined parameters of the association rules, like minimum support and minimum confidence, so that *PB* contains more patterns of select items. A better analysis of select items could be realized by using more patterns. If we wish to study the association between a select item and other frequent items then the exact support values of other items might not be available in *PB*. In this case, the central office sends a request to each branch office to forward the details (like support values) of some items that would be required to study the select items. Hence each branch applies a "traditional" mining algorithm (at interface 3/2) which is completed on its local database and forwards the details of local patterns requested by the central office. Let LPB_i be the details of i-th local pattern base requested by the central office, $i = 1, 2,..., n$. A global pattern mining application of select items might be required to access the local patterns and the patterns in *PB*. A global pattern mining application (interface 6/5) is developed based on the patterns present in *PB* and *LPB*$_i$, $i = 1, 2,..., n$. The technique of mining global patterns of select items is efficient due to the following reasons:

- One could extract more patterns of select items by lowering further the parameters of association rule such as the minimum support and minimum confidence, based on the level of data analysis of select items, since *FD* is reasonably small.
- We get true global patterns of select items as there is no need to estimate them.

In light of these observations, we can anticipate that the quality of global patterns is high, since there is no need to estimate them.

To evaluate the effectiveness of the above technique, we present a problem of multi-database mining. We show how the data mining technique presented above could be used in finding the solution to the problem. We start with the notion of overall association between two items in a database (Adhikari et al. 2011).

6.3 Overall Association Between Two Items in a Database

Let *I(DB)* be the set of items in database *DB*. A common measure of similarity (Wu et al. 2005; Xin et al. 2005) between two objects could be used as a measure of positive association between two items in a database. We define positive association between two items in a database as follows:

$$PA(x, y, DB) = \frac{\# \textit{transaction} \text{ containing both } x \text{ and } y, DB}{\# \text{ transaction containing at least one of } x \text{ and } y, DB},$$

$$\text{for } x, y \in I(DB)$$

where, $\# P, DB$ is the number of transactions in *DB* that satisfy predicate *P*. *PA* measures only positive association present between two items in a database. It does not measure negative association between two items in a database. In the following example, we show that *PA* fails to compute an overall association between two items.

Example 6.1 Let us consider four branches of a multi-branch company. Let D_i be the database corresponding to the *i*-th branch of the company, $i = 1, 2, 3, 4$. The company is interested in analyzing globally a set of select items (*SI*). Let $SI = \{a, b\}$. The content of different databases is given as follows: $D_1 = \{\{a, e\}, \{b, c, g\}, \{b, e, f\}, \{g, i\}\}$; $D_2 = \{\{b, c\}, \{f, h\}\}$; $D_3 = \{\{a, b, c\}, \{a, e\}, \{c, d\}, \{g\}\}$; $D_4 = \{\{a, e\}, \{b, c, g\}\}$. Initially, we wish to measure the association between two items in a single database, say D_1. Now, $PA(a, b, D_1) = 0$, since there is no transaction in D_1 containing both the items *a* and *b*. In these transactions, if one of the items of $\{a, b\}$ is present then the other item of $\{a, b\}$ is not present. Thus, the transactions $\{a, e\}, \{b, c, g\}$ and $\{b, e, f\}$ in D_1 imply that the items a and b are negatively associated. We need to define a measure of negative association between two items in a database. Similar to the measure of positive association, one defines a measure of negative association between two items in a database as follows:

$$NA(x, y, DB) = \frac{\# \text{ transaction containing exactly one of } x \text{ and } y, DB}{\# \text{ transaction containing at least one of } x \text{ and } y, DB},$$

for $x, y \in I(DB)$.

Now, $NA(a, b, D_1) = 1$. We note that $PA(a, b, D_1) < NA(a, b, D_1)$. Overall, we state that the items a and b are negatively associated, and the level of overall association between the items a and b in D_1 is $PA(a, b, D_1) - NA(a, b, D_1) = -1.0$. The accuracy of association analysis might be low if we consider only the positive association between two items. •

The analysis of relationships among variables is a fundamental task being at the heart of many data mining problems. For example, metrics such as support, confidence, lift, correlation, and collective strength have been used extensively to evaluate the interestingness of association patterns (Klemettinen et al. 1994; Silberschatz and Tuzhilin 1996; Aggarwal and Yu 1998; Silverstein et al. 1998; Liu et al. 1999). These metrics are defined in terms of the frequency counts tabulated in a 2×2 contingency table as shown in Table 6.1. Tan et al. (2002) presented an overview of twenty one interestingness measures proposed in statistics, machine learning and data mining literature. We continue our discussion with the examples cited in Tan et al. (2002) and show that none of the proposed measures is effective in finding the overall association by considering both positive and negative associations between two items in a database.

From the examples shown in Table 6.2, we notice that the overall association level between two items could be negative as well as positive. In fact, a measure of overall association between two items in a database produces results in $[-1, 1]$. We consider the following five out of twenty one interestingness measures, since the association between two items calculated using one of these five measures lies in $[-1, 1]$. Thus, we study their usefulness for the specific requirement of the proposed problem. These five measures are included in Table 6.3.

In Table 6.4, we rank the contingency tables by using each of the above measures.

Table 6.1 A 2×2 contingency table for variables x and y		y	$\neg y$	Total
	x	f_{11}	f_{10}	$f_{1.}$
	$\neg x$	f_{01}	f_{00}	$f_{0.}$
	Total	$f_{.1}$	$f_{.0}$	$f_{..}$

Table 6.2 Examples of contingency tables

Example	f_{11}	f_{10}	f_{01}	f_{00}
E1	8,123	83	424	1,370
E2	8,330	2	622	1,046
E3	9,481	94	127	298
E4	3,954	3,080	5	2,961
E5	2,886	1,363	1,320	4,431
E6	1,500	2,000	500	6,000
E7	4,000	2,000	1,000	3,000
E8	4,000	2,000	2,000	2,000
E9	1,720	7,121	5	1,154
E10	61	2,483	4	7,452

Table 6.3 Selected interestingness measures for association patterns

Symbol	Measure	Formula
ϕ	ϕ-coefficient	$\dfrac{P(\{x\}\cup\{y\}) - P(\{x\})\times P(\{y\})}{\sqrt{P(\{x\})\times P(\{y\})\times(1-P(\{x\})\times(1-P(\{y\})))}}$
Q	Yule's Q	$\dfrac{P(\{x\}\cup\{y\})\times P(\neg(\{x\}\cap\{y\})) - P(\{x\}\cup\neg\{y\})\times P(\neg\{x\}\cup\{y\})}{P(\{x\}\cup\{y\})\times P(\neg(\{x\}\cap\{y\})) - P(\{x\}\cup\neg\{y\})\times P(\neg\{x\}\cup\{y\})}$
Y	Yule's Y	$\dfrac{\sqrt{P(\{x\}\cup\{y\})\times P(\neg(\{x\}\cap\{y\}))} - \sqrt{P(\{x\}\cup\neg\{y\})\times P(\neg\{x\}\cup\{y\})}}{\sqrt{P(\{x\}\cup\{y\})\times P(\neg(\{x\}\cap\{y\}))} - \sqrt{P(\{x\}\cup\neg\{y\})\times P(\neg\{x\}\cup\{y\})}}$
κ	Cohen's κ	$\dfrac{P(\{x\}\cup\{y\}) + P(\neg\{x\}\cup\neg\{y\}) - P(\{x\})\times P(\{y\}) - P(\neg\{x\})\times P(\neg\{y\})}{1 - P(\{x\})\times P(\{y\}) - P(\neg\{x\})\times P(\neg\{y\})}$
F	Certainty factor	$\max\left(\dfrac{P(\{y\}\mid\{x\}) - P(\{y\})}{1 - P(\{y\})}, \dfrac{P(\{x\}\mid\{y\}) - P(\{x\})}{1 - P(\{x\})}\right)$

Table 6.4 Ranking of contingency tables using above interestingness measures

Example	E1	E2	E3	E4	E5	E6	E7	E8	E9	E10
ϕ	1	2	3	4	5	6	7	8	9	10
Q	3	1	4	2	8	7	9	10	5	6
Y	3	1	4	2	8	7	9	10	5	6
κ	1	2	3	5	4	7	6	8	9	10
F	4	1	6	2	9	7	8	10	3	5

Also, we rank the contingency tables based on the concept of overall association explained in Example 6.1. In Table 6.5, we present the ranking of contingency tables using overall association.

The ranks given in Table 6.5 and the ranks given for each of the five measures in Table 6.4 are not similar. In other words, none of the above five measures satisfies the requirement formulated in the proposed problem. Based on the above discussion, we propose the following measure *OA* as an overall association between two items in a database.

Table 6.5 Ranking of contingency tables using overall association

Example	Overall association	Rank
E1	0.76	3
E2	0.77	2
E3	0.93	1
E4	0.09	5
E5	0.02	6
E6	−0.10	8
E7	0.10	4
E8	0	7
E9	−0.54	10
E10	−0.24	9

Definition 6.1 $OA(x, y, DB) = PA(x, y, DB) - NA(x, y, DB)$, for $x, y \in I(DB)$. •

If $OA(x, y, DB) > 0$ then the items x and y are positively associated in DB. If $OA(x, y, DB) < 0$ then the items x and y are negatively associated in DB. The problem is concerned with the association between a nucleus item and another item in a database. Thus, we are not concerned about the association between two items in a group, where none of them is a nucleus item. In other words, it could be considered as a problem of grouping rather than a problem of classification or clustering.

6.4 An Application: Study of Select Items in Multiple Databases Through Grouping

As before, let us consider a multi-branch company having n branches. Each branch maintains a separate database for the transactions made in that particular branch. Let D_i be the database corresponding to the i-th branch of the multi-branch company, $i = 1, 2,..., n$. Also, let D be the union of all branch databases. A large section of a local database might be irrelevant to the current problem. Thus, we divide database D_i into FD_i and RD_i, where FD_i and RD_i are called the *forwarded database* and *remaining database* corresponding to the i-th branch, respectively, $i = 1, 2,..., n$. We are interested in the forwarded databases, since every transaction in a forwarded database contains at least one select item. The database FD_i is forwarded to the central office for mining global patterns of select items, $i = 1, 2,..., n$. All the local forwarded databases are amassed into a single database (FD) for the purpose of mining task. We note that the database FD is not overly large as it contains transactions related to select items. Before proceeding with the detailed discussion, we first offer some definitions.

A set of items is referred to as an *itemset*. An itemset containing k items is called a *k-itemset*. The *support* (*supp*) (Agrawal et al. 1993) of an itemset refers to the fraction of transactions containing this itemset. If an itemset satisfies the user-specified minimum support (α) criterion, then it is called a *frequent itemset* (*FIS*).

Similarly, if an item satisfies the user-specified minimum support criterion, then it is called a *frequent item* (*FI*). If a k-itemset is frequent then every item in the k-itemset is also frequent. In this chapter, we study the items in *SI*. Let $SI = \{s_1, s_2,..., s_m\}$. We wish to construct m groups of frequent items in such a way that the i-th group grows by being centered around the nucleus item s_i, $i = 1, 2,..., m$. Let FD be the union of FD_i, $i = 1, 2,..., n$. Furthermore let $FIS_k(DB \mid \alpha)$ be the set of frequent k-itemsets in database DB at the minimum support level α, $k = 1, 2$. We state our problem as follows:

Let G_i be the i-the group of frequent items containing the nucleus item $s_i \in SI$, $i = 1, 2,..., m$. Construct G_i using $FIS_2(FD \mid \alpha)$ and local patterns in D_i such that $x \in G_i$ implies $OA(s_i, x, D) > 0$, for $i = 1, 2,..., m$.

Two groups may not be mutually exclusive, as our study involves identifying pairs of items such that the following conditions are true: (1) the items in each pair are positively associated between each other in D, and (2) one of the items in a pair is a select item. Our study is not concerned with the association between a pair of items in a group such that none of them is a select item. The above problem actually results in $m + 1$ groups where $(m + 1)$-th group G_{m+1} contains the items that are not positively associated with any one of the select items. The proposed study is not concerned with the items in G_{m+1}.

The crux of the proposed problem is to determine the supports of the relevant frequent itemsets in multiple large databases. A technique of estimating support of a frequent itemset in multiple real databases has been proposed by Adhikari and Rao (2008a). To estimate the support of an itemset in a database, this technique makes use of the trend of supports of the same itemset in other databases. The trend approach for estimating support of an itemset in a database could be described as follows:

Let the itemset X gets reported from databases $D_1, D_2,..., D_m$. Also let *supp* $(X, \cup_{i=1}^{m} D_i)$ be the support of X in the union of $D_1, D_2,..., D_m$. Let D_k be a database that does not report X, for $k = m + 1, m + 2,..., n$. Then the support of X in D_k could be estimated by $\alpha \times supp(X, \cup_{i=1}^{m} D_i)$. Given an itemset X, some local supports of X are estimated and the remaining local supports of X are obtained using a traditional data mining technique. The global support of X is obtained by combining these local supports with the numbers of transactions (i.e., sizes) of the respective databases. The proposed technique synthesizes true supports of relevant frequent itemsets in multiple databases.

In the previous chapters, we have discussed the limitations of the traditional way of mining multiple large databases. We have observed that local pattern analysis alone could not provide an effective solution to this problem. The mining technique visualized in Fig. 6.1 offers a viable solution. A pattern based on all the databases is called a *global pattern*. A global pattern containing at least one select item is called a *global pattern of select item*.

6.4.1 Properties of Different Measures

If the itemset $\{x, y\}$ is frequent in DB then $OA(x, y, DB)$ is not necessarily be positive, since the number of transactions containing only one of the items of $\{x, y\}$ could be more than the number of transactions containing both the items x and y. $OA(x, y, DB)$ could attain maximum value for an infrequent itemset $\{x, y\}$ also. Let $\{x, y\}$ be infrequent. The distributions of x and y in DB are such that no transaction in DB contains only one item of $\{x, y\}$. Thus, $OA(x, y, DB) = 1.0$. In what follows, we discuss a few properties of different measures.

Lemma 6.1 (i) $0 \leq PA(x, y, DB) \leq 1$; (ii) $0 \leq NA(x, y, DB) \leq 1$; (iii) $-1 \leq OA(x, y, DB) \leq 1$; (iv) $PA(x, y, DB) + NA(x, y, DB) = 1$; for $x, y \in I(DB)$. •

$PA(x, y, DB)$ could be considered as a similarity between x and y in DB. Thus, $1 - PA(x, y, DB)$ i.e., $NA(x, y, DB)$ could be considered as a distance between x and y in DB. A characteristic of a good distance measure is that it satisfies metric properties (Barte 1976) over the concerned domain.

Lemma 6.2 $NA(x, y, DB) = 1 - PA(x, y, DB)$ is a metric over $[0, 1]$, for $x, y \in I(DB)$. •

Proof We prove only the property of triangular inequality, since the remaining two properties of the metric are obvious. Let $I(DB) = \{a_1, a_2, ..., a_N\}$. Let ST_i be the set of transactions containing item $a_i \in I(DB)$, $i = 1, 2, ..., N$.

$$1 - PA(a_p, a_q, DB) = 1 - \frac{|ST_p \cap ST_q|}{|ST_p \cup ST_q|} \geq \frac{|ST_p - ST_q| + |ST_q - ST_p|}{|ST_p \cup ST_q \cup ST_r|} \quad (6.3)$$

Thus,

$$1 - PA(a_p, a_q, DB) + 1 \\ - PA(a_q, a_r, DB) \geq \frac{|ST_p - ST_q| + |ST_q - ST_p| + |ST_q - ST_r| + |ST_r - ST_q|}{|ST_p \cup ST_q \cup ST_r|} \quad (6.4)$$

$$= \frac{|ST_p \cup ST_q \cup ST_r| - |ST_p \cap ST_q \cap ST_r| + |ST_p \cap ST_r| + |ST_q| - |ST_p \cap ST_q| - |ST_q \cap ST_r|}{|ST_p \cup ST_q \cup ST_r|} \quad (6.5)$$

$$= 1 - \frac{|ST_p \cap ST_q \cap ST_s| - |ST_p \cap ST_s| - |ST_q| + |ST_p \cap ST_q| + |ST_q \cap ST_s|}{|ST_p \cup ST_q \cup ST_s|} \quad (6.6)$$

$$= 1 - \frac{\{ |ST_p \cap ST_q \cap ST_s| + |ST_p \cap ST_q| + |ST_q \cap ST_s| \} - \{ |ST_p \cap ST_s| + |ST_q| \}}{|ST_p \cup ST_q \cup ST_s|} \quad (6.7)$$

Let the number of elements in the shaded region of Figs. 6.2c and d be N_1 and N_2, respectively. Then the expression (6.7) becomes

$$1 - \frac{N_1 - N_2}{|ST_p \cup ST_q \cup ST_r|} \geq \begin{cases} 1 - \frac{N_1 - N_2}{|ST_p \cup ST_q \cup ST_r|}, & \text{if } N_1 \geq N_2 \ (\text{case 1}) \\ 1 - \frac{|ST_p \cap ST_r|}{|ST_p \cup ST_q \cup ST_r|}, & \text{if } N_1 < N_2 \ (\text{case 2}) \end{cases} \quad (6.8)$$

In case 1, the expression remains the same. In case 2, a positive quantity $ST_p \cap ST_r$ has been put in place of a negative quantity N_1–N_2. Thus the expression (6.8) reads as

$$\geq \begin{cases} 1 - \frac{N_1 - N_2}{|ST_p \cup ST_r|}, & \text{if } N_1 \geq N_2 \\ 1 - \frac{|ST_p \cap ST_r|}{|ST_p \cup ST_r|}, & \text{if } N_1 < N_2 \end{cases} \geq \begin{cases} 1 - \frac{N_1}{|ST_p \cup ST_r|}, & \text{if } N_1 \geq N_2 \\ 1 - \frac{|ST_p \cap ST_r|}{|ST_p \cup ST_r|}, & \text{if } N_1 < N_2 \end{cases}$$
$$\geq \begin{cases} 1 - \frac{|ST_p \cap ST_r|}{|ST_p \cup ST_r|}, & \text{if } N_1 \geq N_2 \\ 1 - \frac{|ST_p \cap ST_r|}{|ST_p \cup ST_r|}, & \text{if } N_1 < N_2 \end{cases} \quad (6.9)$$

where $N_1 = |ST_p \cap ST_q \cap ST_r| \leq |ST_p \cap ST_r|$. Therefore, irrespectively of the relationship between N_1 and N_2, $1-PA(a_p, a_q, DB) + 1-PA(a_q, a_r, DB) \geq 1-PA$ (a_p, a_r, DB). Thus, $1-PA(x, y, DB)$ satisfies the requirement of triangular inequality. •

To compute an overall association between two items, we need to express OA in terms of supports of frequent itemsets.

Lemma 6.3 *For any two items $x, y \in I(DB)$, $OA(x, y, DB)$ can be expressed as follows:*

$$OA(x, y, DB) = \frac{3 \times supp(\{x, y\}, DB) - supp(\{x\}, DB) - supp(\{y\}, DB)}{supp(\{x\}, DB) + supp(\{y\}, DB) - supp(\{x, y\}, DB)}$$
$$(6.10)$$

Proof $OA(x, y, DB) = PA(x, y, DB)-NA(x, y, DB)$

Now, $PA(x, y, DB) = \dfrac{supp(\{x, y\}, DB)}{supp(\{x\}, DB) + supp(\{y\}, DB) - supp(\{x, y\}, DB)}$
$$(6.11)$$

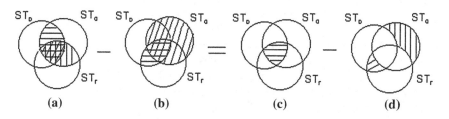

Fig. 6.2 Simplification using Venn diagram

Also, $NA(x, y, DB) = \dfrac{supp(\{x\}, DB) + supp(\{y\}, DB) - 2 \times supp(\{x, y\}, DB)}{supp(\{x\}, DB) + supp(\{y\}, DB) - supp(\{x, y\}, DB)}$

$$(6.12)$$

Thus, the lemma follows. •

6.4.2 Grouping Frequent Items

For the purpose of explaining the grouping process, we continue our discussion of Example 6.1.

Example 6.2 Based on *SI*, the forwarded databases are given as follows:

$FD_1 = \{\{a, e\}, \{b, c, g\}, \{b, e, f\}\}$

$FD_2 = \{\{b, c\}\}$

$FD_3 = \{\{a, b, c\}, \{a, e\}\}$

$FD_4 = \{\{a, e\}, \{b, c, g\}\}$

Let *size(DB)* be the number of transactions in *DB*. Then $size(D_1) = 4$, $size(D_2) = 2$, $size(D_3) = 4$, and $size(D_4) = 2$. The union of all forwarded databases is given as

$FD = \{\{a, e\}, \{b, c, g\}, \{b, e, f\}, \{b, c\}, \{a, b, c\}, \{a, e\}, \{a, e\}, \{b, c, g\}\}$.

The transaction $\{a, e\}$ has been shown three times, since it has originated from three data sources. We mine the database *FD* and get the following set of frequent itemsets:

$FIS_1(FD \mid 1/14) = \{\{a\}\ (4/12), \{b\}\ (5/12)\}$

$FIS_2(FD \mid 1/14) = \{\{a, b\}\ (1/12), \{a, c\}\ (1/12), \{a, e\}\ (3/12), \{b, c\}\ (4/12), \{b, e\}\ (1/12), \{b, f\}\ (1/12), \{b, g\}\ (2/12)\}$

where $X(\eta)$ denotes the fact that the frequent itemset X has support η. All the transactions containing item x not belonging to *SI* might not be available in *FD*. Thus other frequent itemsets of size one could not be mined correctly from *FD*. They are not shown in $FIS_1(FD)$. Each frequent itemset extracted from *FD* contains an item from *SI*. The collection of patterns in $FIS_1(FD \mid 1/14)$ and $FIS_2(FD \mid 1/14)$ could be considered as *PB* with reference to Fig. 6.1. Using the frequent itemsets in $FIS_1(FD \mid \alpha)$ and $FIS_2(FD \mid \alpha)$ we might not be able to compute the value of *OA* between two items. The central office of the company requests each branch for the supports of the relevant items (*RIs*) to calculate the overall association between two items. Such information would help the central office to compute exactly the value of the overall association in the union of all databases. Relevant items are the items in $FIS_1(FD \mid \alpha)$ that do not belong to *SI*. In this example, *RIs* are c, e, f and g. The supports of relevant items in different databases are given below:

$RI(D_1) = \{\{c\}\ (1/4), \{e\}\ (2/4), \{f\}\ (1/4), \{g\}\ (2/4)\}$

$RI(D_2) = \{\{c\}\ (1/2), \{e\}\ (0), \{f\}\ (1/2), \{g\}\ (0)\}$

$RI(D_3) = \{\{c\}(2/4), \{e\}(1/4), \{f\}(0), \{g\}(1/4)\}$

$RI(D_4) = \{\{c\}\ (1/2), \{e\}\ (1/2), \{f\}\ (0), \{g\}\ (1/2)\}$ •

$RI(D_i)$ could be considered as LPB_i with reference to Fig. 6.1, $i = 1, 2,..., n$. We follow here a grouping technique based on the proposed measure of overall association OA. If $OA(x, y, D) > 0$ then y could be placed in the group of x, for $x \in SI = \{a, b\}, y \in I(D)$. We explain the procedure of grouping frequent items with the help of following example.

Example 6.3 Here we continue the discussion of Example 6.2. Based on the available supports of local 1-itemsets, we synthesize 1-itemsets in D as mentioned in Table 6.6.

We note that the supports of $\{a\}$ and $\{b\}$ are not required to be synthesized, since they could be determined exactly from mining FD. The values of OA corresponding to itemsets of FIS_2 are presented in Table 6.7.

In Table 6.7, we find that the items a and e are positively associated. Thus, item e could be placed in the group containing nucleus item a. Items b and c are positively associated as well. Item c could be put in the group containing nucleus item b. Thus, the output grouping π using the proposed technique comes in the form:

$$\pi(FIS_1(D)|\{a, b\}, 1/12) = \{\text{Group 1}, \text{Group 2}\},$$

where

Group 1 $= \{(a, 1.0), (e, 0.2)\}$
Group 2 $= \{(b, 0.1), (c, 0.33)\}$.

Each item in a group is associated with a real number, which represents the strength of an overall association between the item and the nucleus item of the group. Using The proposed grouping technique we also construct the third group of items, i.e., $\{f, g\}$. The proposed study is not concerned with the items in $\{f, g\}$. •

Each group grows being centered around a select item. The i-th group (G_i) grows centering around the i-th select item s_i, $i = 1, 2,..., m$. With respect to group G_i, the item s_i is called the nucleus item of G_i, $i = 1, 2,..., m$. We define a group as follows.

Definition 6.2 The i-th group is a collection of frequent items a_j and the nucleus item $s_i \in SI$ such that $OA(s_i, a_j, D) > 0, j = 1, 2,..., |G_i|$, and $i = 1, 2,..., m$. •

Table 6.6 Supports of relevant 1-itemsets in D

Itemset ($\{x\}$)	$\{a\}$	$\{b\}$	$\{c\}$	$\{e\}$	$\{f\}$	$\{g\}$
$supp(\{x\}, D)$	4/12	5/12	5/12	4/12	2/12	4/12

Table 6.7 Overall association between two items in a frequent 2-itemset in FD

Itemset ($\{x, y\}$)	$\{a, b\}$	$\{a, c\}$	$\{a, e\}$	$\{b, c\}$	$\{b, e\}$	$\{b, f\}$	$\{b, g\}$
$OA(x, y, D)$	−3/4	−3/4	1/5	1/3	−3/4	−2/3	−3/7

Let us describe the data structures used in the algorithm for finding groups. The set of frequent k-itemsets is maintained in an array *FISk*, $k = 1, 2$. After finding *OA* value between two items in a 2-itemset, it is kept in array *IS2*. Thus, the number of itemsets in *IS2* is equal to the number of frequent 2-itemsets extracted from *FD*. A two-dimensional array *Groups* is maintained to store m groups. The i-the row of *Groups* stores the i-th group, for $i = 1, 2,..., m$. The first element of i-th row contains the i-th select item, for $i = 1, 2,..., m$. In general, the j-th element of the i-th row contains a pair (*item, value*), where *item* refers to the j-th item of the i-th group and *value* refers to the amount of *OA* value between the i-th select item and *item*, for $j = 1, 2,..., |G_i|$. The grouping algorithm can be outlined as follows.

Algorithm 6.1 Construct m groups of frequent items in D such that i-th group grows being centered around the i-th select item, for $i = 1, 2, ..., m$.
procedure m-grouping ($m, SI, N_1, FIS1, N_2, FIS2, GSize, Groups$)

    Input: $m, SI, N_1, FIS1, N_2, FIS2$
    m: the number of select items
    SI: set of select items
    N_k: number of frequent k-itemsets
    FISk: set of frequent k-itemsets
    Output: *GSize, Groups*
    GSize: array of number of elements in each group
    Groups: array of m groups
    01: **for** $i = 1$ to N_2 **do**
    02:     $IS2(i).value = OA(FIS2(i).item1, FIS2(i).item2, D)$;
    03:     $IS2(i).item1 = FIS2(i).item1$; $IS2(i).item2 = FIS2(i).item2$;
    04: **end for**
    05: **for** $i = 1$ to m **do**
    06:    $Groups(i)(1).item = SI(i)$; $Groups(i)(1).value = 1.0$; $GSize(i) = 1$;
    07: **end for**
    08: **for** $i = 1$ to N_2 **do**
    09:    **for** $j = 1$ to m **do**
    10:     **if** (($IS2(i).item1 = SI(j)$) **and** ($IS2(i).value > 0$)) **then**
    11:      $GSize(j) = GSize(j) + 1$; $Groups(j)(GSize(j)).item = IS2(i).item2$;
    12:      $Groups(j)(GSize(j)).value = IS2(i).value$;
    13:     **end if**
    14:     **if** (($IS2(i).item2 = SI(j)$) **and** ($IS2(i).value > 0$)) **then**
    15:      $GSize(j) = GSize(j) + 1$; $Groups(j)(GSize(j)).item = IS2(i).item1$;
    16:      $Groups(j)(GSize(j)).value = IS2(i).value$;
    17:     **end if**
    18:    **end for**
    19: **end for**
    20: **for** $i = 1$ to m **do**
    21:   sort items of group i in non-increasing order on *OA* value;
    22: **end for**
    end procedure

The algorithm works as follows. Using (6.10), we compute the value of *OA* for each itemset in *FIS2*. After computing *OA* value for a pair of items, we store the items and *OA* value in *IS2*. The algorithm performs these tasks using the for-loop shown in lines 01–04. We initialize each group with the corresponding nucleus item as shown in lines 05–07. A relevant item or an item in *SI* could belong to one or more groups. Thus, we check for the possibility of including each of the relevant items and items in *SI* to each group using the for-loop (lines 09–18). All the relevant items and items in *SI* are covered using for-loop present in lines 08–19. For the purpose of better presentation, we finally sort items of i-th group in non-increasing order on *OA* value, $i = 1, 2,..., m$.

Assume that the frequent itemsets in *FIS1* and *FIS2* are sorted on items in the itemset. Thus, the time complexities for searching an itemset in *FIS1* and *FIS2* are $O(log(N_1))$ and $O(log(N_2))$, respectively. The time complexity of computing present at line 02 is $O(log(N_1))$, since $N_1 > N_2$. The time complexity of calculations carried out in lines 01–04 is $O(N_2 \times log(N_1))$. Lines 05–07 are used to complete all necessary initialization. The time complexity of this program segment is $O(m)$. Lines 08–19 process frequent 2-itemsets and construct m groups. If one of the two items in a frequent 2-itemset is a select item, then other item could be placed in the group of the select item, provided the overall association between them is positive. The time complexity of this program segment is $O(m \times N_2)$. Lines 20–22 present groups in sorted order. Each group is sorted in non-increasing order with respect to the *OA* value. The association of nucleus item with itself is 1.0. Thus the nucleus item is kept at the beginning of the group (line 06). Let the average size of a group be k. Then the time complexity of this program segment is $O(m \times k \times log(k))$. The time complexity of the procedure *m-grouping* is *maximum* $\{O(N_2 \times log(N_1)), O(m), O(m \times N_2), O(m \times k \times log(k))\}$, i.e., *maximum* $\{O(N_2 \times log(N_1)), O(m \times N_2), O(m \times k \times log(k))\}$.

6.4.3 Experiments

We have carried out a suite of experiments to quantify the effectiveness of the above approach. We present the experimental results using four databases, viz., *retail* (Frequent itemset mining dataset repository 2004), *mushroom* (Frequent itemset mining dataset repository 2004), *T10I4D100K* (Frequent itemset mining dataset repository 2004), and *check*. The database *retail* obtained from an anonymous Belgian retail supermarket store and concerns a real-world problem. The database *mushroom* comes from the UCI databases. The database *T10I4D100K* is synthetic and was obtained using a generator from IBM Almaden Quest research group. The database *check* is artificial whose grouping is already known. We have experimented with database *check* to verify that our grouping technique works correctly. We present some characteristics of these databases in Table 6.8. Let *NT*, *AFI*, *ALT*, and *NI* denote the number of transactions, average frequency of an item, average length of a transaction, and number of items in the database, respectively.

We divide each of these databases into ten databases called here *input databases*. The input databases obtained from *R*, *M*, *T* and *C* are names as R_i, M_i, T_i, and C_i, respectively, $i = 1, 2,..., 10$. We present some characteristics of the input databases in Tables 6.9 and 6.10.

Table 6.8 Characteristics of databases used in the experiment

Database	NT	ALT	AFI	NI
Retail (R)	88,162	11.31	99.67	10,000
Mushroom (M)	8,124	24.00	1624.80	120
T10I4D100K (T)	1,00,000	11.10	1276.12	870
Check (C)	40	3.03	3.10	39

Table 6.9 Characteristics of input databases obtained from *retail* and *mushroom*

DB	NT	ALT	AFI	NI	DB	NT	ALT	AFI	NI
R_1	9,000	11.24	12.07	8,384	M_1	812	24.00	295.27	66
R_2	9,000	11.21	12.27	8,225	M_2	812	24.00	286.59	68
R_3	9,000	11.34	14.60	6,990	M_3	812	24.00	249.85	78
R_4	9,000	11.49	16.66	6,206	M_4	812	24.00	282.43	69
R_5	9,000	10.96	16.04	6,148	M_5	812	24.00	259.84	75
R_6	9,000	10.86	16.71	5,847	M_6	812	24.00	221.45	88
R_7	9,000	11.20	17.42	5,788	M_7	812	24.00	216.53	90
R_8	9,000	11.16	17.35	5,788	M_8	812	24.00	191.06	102
R_9	9,000	12.00	18.69	5,777	M_9	812	24.00	229.27	85
R_{10}	9,000	11.69	15.35	5,456	M_{10}	816	24.00	227.72	86

Table 6.10 (continued). Characteristics of input databases obtained from *T10I4D100K*

DB	ALT	AFI	NI	DB	ALT	AFI	NI
T_1	11.06	127.66	866	T_6	11.14	128.63	866
T_2	11.13	128.41	867	T_7	11.11	128.56	864
T_3	11.07	127.65	867	T_8	11.10	128.45	864
T_4	11.12	128.44	866	T_9	11.08	128.56	862
T_5	11.13	128.75	865	T_{10}	11.08	128.11	865

The input databases obtained from database *check* are given as follows:

$C_1 = \{\{1, 4, 9, 31\}, \{2, 3, 44, 50\}, \{6, 15, 19\}, \{30, 32, 42\}\}$
$C_2 = \{\{1, 4, 7, 10, 50\}, \{3, 44\}, \{11, 21, 49\}, \{41, 45, 59\}\}$
$C_3 = \{\{1, 4, 10, 20, 24\}, \{5,7, 21\}, \{21, 24, 39\}, \{26, 41, 46\}\}$
$C_4 = \{\{1, 4, 10, 23\}, \{5, 8\}, \{5, 11, 21\}, \{42, 47\}\}$
$C_5 = \{\{1, 4, 10, 34\}, \{5, 49\}, \{25, 39, 49\}, \{49\}\}$
$C_6 = \{\{1, 3, 44\}, \{6, 41\}, \{22, 26, 38\}, \{45, 49\}\}$
$C_7 = \{\{1, 2, 3, 10, 20, 44\}, \{11, 12, 13\}, \{24, 35\}, \{47, 48, 49\}\}$
$C_8 = \{\{2, 3, 20, 39\}, \{2, 3, 20, 44, 50\}, \{32, 49\}, \{42, 45\}\}$
$C_9 = \{\{2, 3, 20, 44\}, \{3, 19, 50\}, \{5, 41, 45\}, \{21\}\}$
$C_{10} = \{\{2, 20, 45\}, \{5, 7, 21\}, \{11, 19\}, \{22, 30, 31\}\}$

In Table 6.11, we present some relevant details regarding different experiments. We have chosen the first 10 frequent items as the select items, except for the last

Table 6.11 Some relevant information regarding experiments

Database	α	SI
R	0.03	{0,1,2,3,4,5,6,7,8, 9}
M	0.05	{1, 3, 9, 13, 23, 34, 36, 38, 40, 52}
T	0.01	{2, 25, 52, 240, 274, 368, 448, 538, 561, 630}
C	0.07	{1, 2, 3}

experiment. One could choose select items as the items whose data analyses are needed to be performed.

The first experiment is based on database *retail*. The grouping of frequent items in *retail* is given below:

π (*FI*(*retail*) | *SI*, α) = { $\underline{0}$ (1.00); $\underline{1}$ (1.00); $\underline{2}$ (1.00); $\underline{3}$ (1.00); $\underline{4}$ (1.00); $\underline{5}$ (1.00); $\underline{6}$ (1.00); $\underline{7}$ (1.00); $\underline{8}$ (1.00); $\underline{9}$ (1.00) }

Two resulting groups are separated by semicolon (;). The nucleus item in each group is underlined. Each item in a group is associated with a real number shown in bracket. This value represents the strength of the overall association between the item and the nucleus item. The groups are shaded for the purpose of clarity of visualization. We observe that no item in database *retail* is positively associated with the select items using the measure *OA*. This does not necessarily mean that the level of AE or ME for the experiment is zero. There may exist frequent itemsets of size two such that overall association between two items in each of the itemsets is non-positive and at least one of the two items belongs to the set of select items.

The second experiment is based on database *mushroom*. The grouping of frequent items in *mushroom* is given below:

π (*FI*(*mushroom*) | *SI*, α) = { $\underline{1}$ (1.00), 24 (0.23, 110 (0.12), 29 (0.10), 36 (0.10), 61 (0.10), 38 (0.06), 66 (0.06), 90 (0.01); $\underline{3}$ (1.00); $\underline{9}$ (1.000000); $\underline{13}$ (1.00); $\underline{23}$ (1.00), 93 (0.53), 59 (0.22), 2 (0.14), 39 (0.01), 63 (0.15); $\underline{34}$ (1.00), 86 (0.99), 85 (0.95), 90 (0.80), 36 (0.63), 39 (0.33), 59 (0.23), 63 (0.17), 53 (0.16), 67 (0.13), 24 (0.12), 76 (0.11); $\underline{36}$ (1.00), 85 (0.68), 90 (0.65), 86 (0.63), 34 (0.63), 59 (0.17), 39 (0.16), 63 (0.11), 110 (0.10), 1 (0.10); $\underline{38}$ (1.00), 48 (0.38), 102 (0.19), 58 (0.14), 1 (0.06), 94 (0.05), 110 (0.01); $\underline{40}$ (1.00); $\underline{52}$ (1.00) }

We observe that some frequent items are not included in any of these groups, since their overall associations with each of the select items are non-positive.

The third experiment is based on database *T10I4D100K*. The grouping of frequent items in *T10I4D100K* is given below:

π (*FI*(*T10I4D100K*) | *SI*, α) = { $\underline{2}$ (1.00); $\underline{25}$ (1.00); $\underline{52}$ (1.00); $\underline{240}$ (1.00); $\underline{274}$ (1.00); $\underline{368}$ (1.00); $\underline{448}$ (1.00); $\underline{538}$ (1.00); $\underline{561}$ (1.00); $\underline{630}$ (1.00) }

We observe that databases *retail* and *T10I4D100K* are sparse. Thus, the grouping contains groups of singleton item for these two databases. The overall association between a nucleus item and itself is 1.0. Otherwise, the overall association between a frequent item and a nucleus item is non-positive for these two databases.

The fourth experiment is based on database *check*. This database is constructed artificially to verify the following existing grouping.

π (*FI(check)* | *SI*, α) = { (1, 1.00), (4, 0.43), (10, 0.43); (2, 1.00), (20, 0.43), (3, 0.11); (3, 1.00), (44, 0.50), (2, 0.11) }.

We have calculated average errors using both trend and the proposed approaches. Figures 6.3, 6.4 and 6.5 show the graphs of AE versus the number databases for the first three databases. The proposed model enables us to find actual supports of all the relevant itemsets in a database. Thus, the AE of an experiment for the proposed approach remains 0. As the number of databases increases, the relative presence of a frequent itemset normally decreases, The error of synthesizing an itemset also increases. Overall, the AE of the experiment using trend approach is likely to increase as the number of databases increases. We observe this phenomenon in Figs. 6.3, 6.4 and 6.5.

Fig. 6.3 AE versus the number of the databases from *retail*

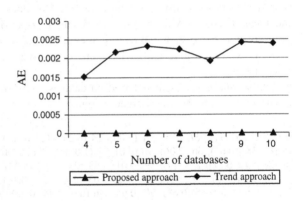

Fig. 6.4 AE versus the number of databases from *mushroom*

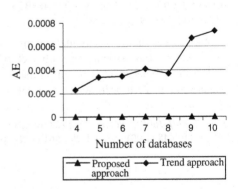

Fig. 6.5 AE versus the
number of databases from
T10I4D100K

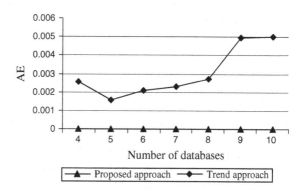

6.5 Related Work

Recently, multi-database mining has been recognized as an important and timely research area in the KDD community. The work reported so far could be classified broadly into two categories: mining/synthesizing patterns in multiple databases and post processing of local patterns. We mention some work related to the first category. Wu and Zhang (2003) have proposed a weighting method for synthesizing high-frequency rules in multiple databases. Zhang et al. (2004a) have developed an algorithm to identify global exceptional patterns in multiple databases. When it comes to the second category, Wu et al. (2005) have proposed a technique for classifying multiple databases for multi-database mining. Using local patterns, we have proposed an efficient technique for clustering multiple databases (Adhikari and Rao 2008b). Lin et al. (2013) have introduced the notion of stable items based on minimum support and variance. Authors have provided a measure of similarity between stable items based on gray relational analysis, and presented a hierarchical gray clustering method for mining stable patterns.

In the context of estimating support of itemsets in a database, Jaroszewicz and Simovici (2002) have proposed a method using Bonferroni-type inequalities (Galambos and Simonelli 1996). The maximum-entropy approach to support estimation of a general Boolean expression is proposed by Pavlov et al. (2000). But these support estimation techniques are suitable for a single database only.

Zhang et al. (2004b), Zhang (2002) have studied various strategies for mining multiple databases. Proefschrift (2004) has studied data mining on multiple relational databases.

Existing parallel mining techniques (Agrawal and Shafer 1999; Chattratichat et al. 1997; Cheung et al. 1996) could also be used to deal with multi-databases. These techniques, however, might provide expensive solutions to studying select items in multiple databases.

6.6 Conclusions

The proposed measure of overall association OA is effective as it considers both positive and negative association between two items. Association analysis of select items in multiple market basket databases is an important as well as highly promising issue, since many data analyses of a multi-branch company are based on select items. One could also apply one of the multi-database mining techniques discussed in Chap. 4. Each technique, except partition algorithm, returns approximate global patterns. On the other hand, the partition algorithm scans each database twice. Therefore, the proposed model of mining global patterns of select items from multiple databases is efficient, since one does not need to estimate the patterns in multiple databases. Moreover, it does not fully scan each database two times.

References

Adhikari A, Rao PR (2008a) Synthesizing heavy association rules from different real data sources. Pattern Recogn Lett 29(1):59–71

Adhikari A, Rao PR (2008b) Efficient clustering of databases induced by local patterns. Decis Support Syst 44(4):925–943

Adhikari A, Ramachandrarao P, Pedrycz W (2011) Study of select items in different data sources by grouping. Knowl Inf Syst 27(1):23–43

Aggarwal C, Yu P (1998) A new framework for itemset generation. In: Proceedings of the 17th symposium on principles of database systems, pp 18–24

Agrawal R, Shafer J (1999) Parallel mining of association rules. IEEE Trans Knowl Data Eng 8(6):962–969

Agrawal R, Imielinski T, Swami A (1993) Mining association rules between sets of items in large databases. In: Proceedings of ACM SIGMOD conference, pp 207–216

Barte RG (1976) The elements of real analysis, 2nd edn. Wiley, London

Chattratichat J, Darlington J, Ghanem M, Guo Y, Hüning H, Köhler M, Sutiwaraphun J, To HW, Yang D (1997) Large scale data mining: challenges, and responses. In: Proceedings of the third international conference on knowledge discovery and data mining, pp 143–146

Cheung D, Ng V, Fu A, Fu Y (1996) Efficient mining of association rules in distributed databases. IEEE Trans Knowl Data Eng 8(6):911–922

Frequent itemset mining dataset repository (2004) http://fimi.cs.helsinki.fi/data

Galambos J, Simonelli I (1996) Bonferroni-type inequalities with applications. Springer, New York

Jaroszewicz S, Simovici DA (2002) Support approximations using Bonferroni-type inequalities. In: Proceedings of sixth European conference on principles of data mining and knowledge discovery, pp 212–223

Klemettinen M, Mannila H, Ronkainen P, Toivonen T, Verkamo A (1994) Finding interesting rules from large sets of discovered association rules. In: Proceedings of the 3rd international conference on information and knowledge management, pp 401–407

Lin Y, Hu X, Li X, Wu X (2013) Mining stable patterns in multiple correlated databases. Decision Support Systems

Liu B, Hsu W, Ma Y (1999) Pruning and summarizing the discovered associations. In: Proceedings of the 5th international conference on knowledge discovery and data mining, pp 125–134

Pavlov D, Mannila H, Smyth P (2000) Probabilistics models for query approximation with large sparse binary data sets. In: Proceedings of sixteenth conference on uncertainty in artificial intelligence, pp 465–472

Proefschrift (2004) Multi-relational data mining, PhD thesis, Dutch Graduate School for Information and Knowledge Systems, Aan de Universiteit Utrecht

Pyle D (1999) Data preparation for data mining. Morgan Kufmann, San Francisco

Silberschatz A, Tuzhilin A (1996) What makes patterns interesting in knowledge discovery systems. IEEE Trans Knowl Data Eng 8(6):970–974

Silverstein C, Brin S, Motwani R (1998) Beyond market baskets: generalizing association rules to dependence rules. Data Min Knowl Disc 2(1):39–68

Tan P-N, Kumar V, Srivastava J (2002) Selecting the right interestingness measure for association patterns. In: Proceedings of SIGKDD conference, pp 32–41

Wu X, Zhang S (2003) Synthesizing high-frequency rules from different data sources. IEEE Trans Knowl Data Eng 14(2):353–367

Wu X, Zhang C, Zhang S (2005) Database classification for multi-database mining. Inf Syst 30(1):71–88

Xin D, Han J, Yan X, Cheng H (2005) Mining compressed frequent-pattern sets. In: Proceedings of the 31st VLDB conference, pp 709–720

Zhang S (2002) Knowledge discovery in multi-databases by analyzing local instances. PhD thesis, Deakin University

Zhang S, Wu X, Zhang C (2003) Multi-database mining. IEEE computational intelligence. Bulletin 2(1):5–13

Zhang C, Liu M, Nie W, Zhang S (2004a) Identifying global exceptional patterns in multi-database mining. IEEE Comput Intel Bull 3(1):19–24

Zhang S, Zhang C, Wu X (2004b) Knowledge discovery in multiple databases. Springer, New York

Chapter 7
Synthesizing Global Exceptional Patterns in Different Data Sources

Many large organizations transact from multiple branches. Many of them possess multiple data sources. The number of multi-branch companies as well as the number of branches of a multi-branch company is increasing over time. Thus, it is important and timely to study data mining carried out on multiple data sources. A global exceptional pattern describes interesting individuality and specificity of few branches. Therefore, it is interesting to identify such patterns. In this chapter, the following main contributions are made:

(1) We propose type I and type II global exceptional frequent itemsets in multiple data sources by extending the notion of global exceptional frequent itemset. (2) We propose a notion of exceptional sources for a type II global exceptional frequent itemset. (3) We also propose type I and type II global exceptional association rules in multiple data sources by extending the notion of global exceptional association rule. (4) We propose an algorithm for synthesizing type II global exceptional frequent itemsets.

Experimental results are presented on both artificial and real-world datasets. We also compare the proposed algorithm with the existing algorithm theoretically and experimentally. The experimental results show that the proposed algorithm is effective and displays a promising performance.

7.1 Introduction

Many multi-branch companies transact from different locations. Many of them collect huge amounts of transactional data continuously through their different branches. Due to a growth-oriented and liberal economic policy adopted by many countries across the globe, the number of such companies as well as the number of branches of such a company is increasing over time. In contrast, the most of the previous data mining works are based on a single database. Thus, it is essential and highly motivated to study data mining on multiple databases. Analysis and synthesis of patterns in multiple databases is an important as well as interesting issue.

Based on the number of data sources, the patterns in multiple databases could be classified into the following categories: local patterns, global patterns and

A. Adhikari et al., *Data Analysis and Pattern Recognition in Multiple Databases*, Intelligent Systems Reference Library 61, DOI: 10.1007/978-3-319-03410-2_7, © Springer International Publishing Switzerland 2014

patterns that are neither local nor global. A pattern based on a single database is called a local pattern. Local patterns are useful for local data analysis. But global patterns are based on all the data sources under consideration. They are useful in global data analyses (Adhikari and Rao 2008c; Wu and Zhang 2003) and global decision-making problems (Adhikari and Rao 2008b; Wu et al. 2005a). In many data mining applications we deal with various types of patterns. For example, frequent itemsets (Agrawal et al. 1993), positive associative rules (Agrawal et al. 1993) and conditional patterns (Adhikari and Rao 2008a). There is no fixed set of attributes to describe these patterns, since patterns are of different types. Each type of pattern could be described by a specific set of attributes. In general, it might be difficult to define a pattern in a database by using certain attributes.

Itemset patterns influence heavily the KDD research in following ways: Firstly, many interesting algorithms have been reported on mining itemset patterns in a database (Agrawal and Srikant 1994; Han et al. 2000; Savasere et al. 1995). Secondly, an itemset could be considered as a basic type of pattern in a transactional database, since many patterns are derived from the itemset patterns in a database. Some examples of derived patterns are positive association rule (Agrawal et al. 1993), negative association rule (Wu et al. 2004), high frequency association rule (Wu and Zhang 2003), and heavy association rule (Adhikari and Rao 2008c). Considerable amount of work have been reported on mining/synthesizing such derived patterns in databases. Thirdly, solutions of many problems are based on the analysis of patterns in a database. Such applications (Wu et al. 2005a; Wang et al. 2001; Adhikari and Rao 2008b) process patterns in a database for the purpose of making some decisions. Thus, the mining and analysis of itemset patterns in a database becomes an interesting as well as important issue.

In the context of itemset pattern we discuss a few concepts and notations that are used in this chapter. *Support* (Agrawal et al. 1993) of an itemset X in database D, denoted by $supp(X, D)$, could be defined as the fraction of transactions in D containing all the items of X. In most of the cases, the importance of an itemset is judged by its support. The itemset X is *frequent* in D if $supp(X, D) \geq \alpha$, where α is user defined level of *minimum support*. Let $SFIS(D)$ be the set of frequent itemsets in database D. Frequent itemsets determine major characteristics of a database. Wu et al. (2005b) have proposed a solution of inverse frequent itemset mining. Authors argued that one could efficiently generate a synthetic market basket dataset from the frequent itemsets and their supports. Let X and Y be two itemsets in D. The characteristics of D are revealed more by the pair $(X, supp(X, D))$ than that of $(Y, supp(Y, D))$, if $supp(X, D) > supp(Y, D)$. Thus, it is important to study frequent itemsets more than infrequent itemsets. In this chapter, we propose two types of exceptional patterns in multiple databases based on local frequency itemsets. We propose algorithms for synthesizing such patterns. There are useful applications of exceptional frequency itemsets. For example, a company might plan to collect the feedback of customers for the exceptional products and implement strategies to increase their sales. Also, the company could identify the branches having high sales of the exceptional items. It might plan to manufacture and/or procure such items locally to reduce the transportation cost. The exceptional frequency items would affect many decisions of a multi-branch company.

The first question that comes to our mind is whether a traditional data mining technique could deal with the multiple large databases. To apply a "traditional" data mining technique, we need to amass all the databases together. A single computer may take an unreasonable amount of time to process the entire database. Sometimes it might not be feasible to carry out the mining task. Another solution would be to employ parallel machines. It requires high investment on hardware and software. Moreover, it is difficult to identify local patterns when a mining technique is applied to the entire database. Thus, the traditional data mining techniques are not suitable in this situation. It seems it is a different problem. Hence, it is required to be dealt with it in a different way where a creative solution could help alleviate the problems highlighted above. One could employ the model of local pattern analysis (Zhang et al. 2003) for mining multiple databases. Under this model of mining multiple databases, each branch requires mining the local database using a traditional data mining technique. Afterwards, each branch forwards the pattern base to the central office. Then the central office processes the pattern bases collected from different branches for synthesizing the global patterns, or making decisions related to different problems.

The rest of the chapter is organized as follows. In Sect. 7.2, we discuss and propose different exceptional patterns in multiple databases. Section 7.3 presents the problem formally. We discuss related work in Sect. 7.4. In Sect. 7.5, we present a technique for synthesizing support of an itemset. Section 7.6 presents an algorithm for synthesizing type II global exceptional itemsets in multiple databases. We define two types of error in Sect. 7.7. Experimental results are presented in Sect. 7.8. Finally, we conclude the chapter in Sect. 7.9.

7.2 Exceptional Patterns in Multiple Data Sources

In this section, we propose type I and type II global exceptional frequent itemsets as well as type I and type II global exceptional association rules in multiple databases. We also discuss other exceptional patterns in multiple databases.

Consider a multi-branch company that has n branches, and with the restriction that all the local transactions are stored locally. Let D_i be the database corresponding to the i-th branch, $i = 1, 2, ..., n$. Also, let D be the union of these databases. In the context of multiple databases, one could define global exceptional patterns in two ways: (1) A pattern that has been extracted from the most of the databases, but has a low support in D (Type I exceptional pattern). (2) A pattern that has been extracted from a few databases, but has a high support in D (Type II exceptional pattern). Both the types of exceptional patterns are global in nature, since all the branch databases are considered. In this chapter, we are interested in mining type II exceptional patterns, we will define here formally both types of global exceptional patterns. Zhang et al. (2004c) have proposed a strategy for mining local exceptions in multiple databases. The authors have defined an exceptional pattern as follows:

A pattern p in local instances is an exceptional pattern if EPI(p) $\geq$ minEP,

where *EPI(p)* is an interestingness measure of *p* defined as follows:

$$EPI(p) = \frac{nExtrn(p) - avgNoExtrn}{-avgNoExtrn}, \qquad (7.1)$$

where *nExtrn(p)* and *avgNoExtrn* are the numbers of times *p* gets extracted and the average number of times a pattern gets extracted from different data sources, respectively. *minEP* is the user-defined threshold of the minimal interest degree. Also, the authors have defined interestingness of a pattern in a branch as follows:

A pattern p in i−th branch is of interest if RI_i (p) $\geq$ minEPsup,

where *RI_i(p)* is the interestingness degree of *p* in the *i*-th branch expressed as follows:

$$RI_i(p) = (supp(p, D_i) - \alpha_i)/\alpha_i, \qquad (7.2)$$

where α_i is the minimum support given for mining D_i, $i = 1, 2, ..., n$; *minEPsup* is the user-defined threshold for minimum interest degree.

From the two definitions given above, the following observations are drawn: (1) The definition of exceptional pattern is considered with respect to all the databases. The definition of interestingness of a pattern is considered with respect to a local database. Thus, an exceptional pattern in multiple databases and interestingness of the pattern in a local branch are two different issues. (2) For a pattern *p* in local instances, the authors have shown that $0 < EPI(p) \leq 1$. We take the following example to show that the above property does not always hold. Let there be only four patterns p_1, p_2, p_3, and p_4 in 15 databases. The number of extractions of these patterns are given as follows: $nExtrn(p_1) = 2$, $nExtrn(p_2) = 15$, $nExtrn(p_3) = 4$, $nExtrn(p_4) = 5$. Thus, $avgNoExtrn = 26/4 = 6.5$. $EPI(p_1) = (2 - 6.5)/(-6.5) = 0.69$, $EPI(p_2) = (15 - 6.5)/(-6.5) = -1.31$, $EPI(p_3) = (4 - 6.5)/(-6.5) = 0.38$, $EPI(p_4) = (5 - 6.5)/(-6.5) = 0.23$. Thus, $EPI(p_2) \notin (0, 1]$. (3) An interesting exceptional pattern might not emerge as a global exceptional pattern, since the support of the pattern is not considered in the union of all databases. It is reasonable that an exceptional frequency itemset should be constrained on the number of times it gets extracted, and its support in the union of all databases. Thus, none of the above two definitions, nor the both the definitions together does serve as a definition of exceptional frequency itemset in multiple databases.

Zhang et al. (2004c) have proposed a technique for identifying global exceptional patterns in multiple databases. The authors described global exceptional pattern as follows:

Global exceptional patterns are highly supported by only a few branches, that is to say, these patterns have very high support in these branches and zero support in other branches.

From the above descriptions, we observe the following points:

1. Let there are ten branches of a multi-branch company. A pattern p has very high support in first two databases that have small sizes. Also, p does not get extracted from the remaining databases. According to the above description, p is a global exceptional pattern. We observe that pattern p might not have high support in the union of all databases. Thus, such description does not serve the purpose. Also, it is not necessarily true that a type II exceptional pattern will have zero supports in the remaining databases. Thus, the above description does not describe type II global exceptional patterns in true sense. Also, we observe the following points in *IdentifyExPattern* algorithm (Zhang et al. 2004a) for identifying exceptional patterns in multiple databases. We believe that the size (i.e., the number of transactions) of a database and support of an itemset in the database are two important parameters for determining the presence of an itemset in a database, since the number of transactions containing the itemset X in a database D_1 is equal to $supp(X, D_1) \times size(D_1)$. The algorithm does not consider size of a database to synthesize the global support of a pattern. Global support of a pattern has been synthesized using only supports of the pattern in the individual databases. We take following example to illustrate this issue. Let there be two databases D_1 and D_2, where $size(D_1)$ is significantly greater than $size(D_2)$. At a given α, we assume that pattern p does not get extracted from D_2, and pattern q does not get extracted from D_1. Thus, $supp(p, D_2)$ and $supp(q, D_1)$ both are assumed as 0. Then, $supp(p, D_1 \cup D_2)$ could be synthesized by $[supp(p, D_1) \times size(D_1) + 0 \times size(D_2)]/size(D_1 \cup D_2)$. If $supp(p, D_1) < supp(q, D_2)$ then it might so happen that $supp(p, D_1 \cup D_2) > supp(q, D_1 \cup D_2)$. In particular, let $size(D_1) = 10{,}000$, $size(D_2) = 100$. At $\alpha = 0.05$, let $supp(p, D_1) = 0.1$, $supp(q, D_1) = 0$, $supp(p, D_2) = 0$, and $supp(q, D_2) = 0.2$. We note that $supp(p, D_1) < supp(q, D_2)$. But, $supp(p, D_1 \cup D_2) = 0.099$, and $supp(q, D_1 \cup D_2) = 0.002$. So, $supp(p, D_1 \cup D_2) > supp(q, D_1 \cup D_2)$. Thus, the size of a database is an important parameter to synthesize the support of a pattern in the union of all databases.

2. The algorithm does not identify type II global exceptional patterns correctly in all the situations. For example, let there be 10 similar databases. Assume that the number of times each pattern gets extracted is either 8, or 9, or 10. Thus, these patterns are supported by most of the databases. According to the nature of type II global exceptional patterns, a high voted pattern is not a type II global exceptional pattern. But, the algorithm would report some of them as type II global exceptional patterns.

3. The algorithm returns patterns that have high supports among the patterns that are extracted less than an average number of times. It is reasonable to consider that a type II global exceptional pattern should exhibit the following properties: (a) the support of a type II global exceptional pattern in the union of all databases is greater than or equal to a user-defined threshold, and (b) the number of extractions of a type II global exceptional pattern is less than a user-defined threshold.

The difficulty of synthesizing a type II global exceptional frequency itemsets is that a frequent itemset in a database may not get reported from every data source. Apart from the synthesized support of an itemset in D, the number of extractions of the itemset is an important issue. An itemset may have high frequency or, low frequency, or neither high frequency nor low frequency. In the context of type II global exceptional pattern, we may need to consider only low frequency itemsets. One could arrive to such a conclusion only if there is a predefined threshold of minimum number of extractions. Thus, a type II global exceptional frequency itemset in multiple data sources could be judged against two thresholds, viz., high support and low extraction. Let γ_1 and γ_2 be the thresholds of low extraction of an itemset and high extraction of an itemset respectively, $0 < \gamma_1 < \gamma_2 \leq 1$. Low and high frequency itemsets are defined as follows.

Definition 1a An itemset X has been extracted from k out of n data sources. Then X has low frequency, if $k < n \times \gamma_1$, where γ_1 is the user-defined threshold of low extraction. •

Definition 1b An itemset X has been extracted from k out of n data sources. Then X has high frequency, if $k \geq n \times \gamma_2$, where γ_2 is the user-defined threshold of high extraction. •

Among low frequency itemsets, we will search for type II global exceptional frequent itemsets. An itemset may not get extracted from all the data sources. Sometimes we need to estimate the support of an itemset in a database to synthesize the support of the itemset in D. Let $supp_a(X, D_i)$ be the actual support of an itemset X in D_i, $i = 1, 2, ..., n$. Let μ_1 and μ_2 be the thresholds of low support and high support for an itemset in a database respectively, $0 \leq \mu_1 < \alpha < \mu_2 \leq 1$. For a single database, we define an itemset with high or low support as follows:

Definition 2a Let X be an itemset in database D_i, $i = 1, 2, ..., n$. X possesses high support in D_i if $supp_a(X, D_i) \geq \mu_2$, where μ_2 ($>\alpha$) is the user-defined threshold of high support, $i = 1, 2, ..., n$. •

Definition 2b Let X be an itemset in database D_i, $i = 1, 2, ..., n$. X possesses low support in D_i if $supp_a(X, D_i) < \mu_1$, where μ_1 ($<\alpha$) is the user-defined threshold of low support, $i = 1, 2, ..., n$. •

The method of synthesizing support of an itemset is discussed in Sect. 7.6. Let $supp_s(X, D)$ be the synthesized support of the itemset X in D. For multiple databases, we define an itemset with high or low support as follows:

Definition 3a Let D be the union of all branch databases. An itemset X in D possesses high support if $supp_s(X, D) \geq \mu_2$, where μ_2 is the user-defined threshold of high support. •

Definition 3b Let D be the union of all branch databases. An itemset X in D possesses low support if $supp_s(X, D) < \mu_1$, where μ_1 is the user-defined threshold of low support. •

Based on the concepts stated above, we propose type II global exceptional frequent itemset in D as follows:

Definition 4 Let D be the union of all branch databases. Let X be a frequent itemset in some branch databases. Then X is a type II global exceptional itemset in D if it has low frequency but high support in D. •

In a similar manner, we propose type I global exceptional frequent itemset in D as follows:

Definition 5 Let D be the union of all branch databases. Let X be a frequent itemset in some branch databases. Then X is a type I global exceptional itemset in D if it has high frequency but low support in D. •

Association rule mining has received a lot of attention in KDD community (Zhang and Wu 2011; Zhang and Zhang 2002; Adamo 2001; Wu et al. 2004). It is based on support-confidence framework established by Agrawal et al. (1993). Let I be set of items in database DB. An association rule r has been expressed symbolically as $X \rightarrow Y$, where $X = \{x_1, x_2, ..., x_p\}$, and $Y = \{y_1, y_2, ..., y_q\}$; $x_i, y_j \in I$, $i = 1, 2, ..., p, j = 1, 2, ..., q$. It expresses association between the itemsets X and Y, called the antecedent and consequent of r respectively. The semantics of this implication could be expressed as follows. If the items in the itemset X are purchased by a customer then the items in the itemset Y are likely to be purchased by the same customer at the same time. The interestingness of an association rule could be expressed by its support and confidence. The support and confidence of association rule r in DB could be expressed as follows: $supp_a(r, DB) = supp_a(XY, DB)$, and $conf_a(r, DB) = supp_a(XY, DB)/supp_a(X, DB)$. One needs to deal with synthesized support ($supp_s$) and synthesized confidence ($conf_s$) of an association rule, since actual support and actual confidence of an association rule in multiple databases might not be available. There exist many techniques by which one could synthesize support of an itemset. Using pipelined feedback technique (Adhikari et al. 2010b) one could obtain a good estimate of support of an itemset in multiple databases. An association rule r in database DB is *interesting* if $supp_a(r, DB) \geq$ *minimum support* (α), and $conf_a(r, DB) \geq$ *minimum confidence* (β). Here the parameters α and β are user-defined. We define *size* of database DB as the number of transactions in DB, denoted by $size(DB)$. In the following we present the concept of heavy association rule in multiple databases. First, we present the concept of heavy association rule in a single database. Afterward, we will present the concept of heavy association rule in multiple databases.

Definition 6a An association rule r in database DB is heavy if $supp_a(r, DB) \geq \mu$, and $conf_a(r, DB) \geq v$, where μ ($> \alpha$) and v ($> \beta$) are the user-defined thresholds of high support and high confidence for identifying heavy association rules in DB respectively (Adhikari and Rao 2008c). •

If an association rule is heavy in a local database then it might not be heavy in the union of all databases D. An association rule in D might have different statuses in different local databases. For example, it might a heavy association rule, or an association rule, or absent in a local database. Thus, one needs to synthesize an association rule for determining its overall status in D. A method of synthesizing an association rule is discussed by Adhikari and Rao (2008c). After synthesizing an association rule, we get synthesized support and synthesized confidence of the association rule in D. Now, we present the concept of heavy association rule in D as follows.

Definition 6b Let D be the union of local databases. An association rule r in D is heavy if $supp_s(r, D) \geq \mu$, and $conf_s(r, D) \geq v$, where μ and v are the user-defined thresholds of high support and high confidence for identifying heavy association rules in D respectively (Adhikari and Rao 2008c). •

Apart from synthesized support and synthesized confidence of an association rule, the frequency of an association rule is an important issue in multi-database mining. We define *frequency* of an association rule as the number of extractions of the association rule from different data sources. If an association rule is extracted from k out of n databases then the frequency of the association rule is k, $0 \leq k \leq n$. An association rule may have low frequency or, high frequency or, neither high frequency nor low frequency in multiple data sources. We could arrive in such a conclusion only if we have user-defined thresholds of low frequency (γ_1), and high frequency (γ_2) of an association rule, $0 < \gamma_1 \leq \gamma_2 \leq 1$. Earlier we have discussed with γ_1 and γ_2 while defining type I and type II global exceptional frequent itemsets. In multi-database mining using local pattern analysis, the concepts of high frequency association rule (Wu and Zhang 2003) and low frequency association rule (Adhikari and Rao 2008c) are presented as follows.

Definition 7a Let an association rule be extracted from k out of n data sources. Then the association rule has low frequency if $k < n \times \gamma_1$, where γ_1 is the user-defined threshold value of low frequency. •

Definition 7b Let an association rule be extracted from k out of n data sources. Then the association rule has high frequency if $k \geq n \times \gamma_2$, where γ_2 is the user-defined threshold of high frequency. •

While synthesizing heavy association rules in multiple databases, it may be worth noting the other attributes of a synthesized association rule. For example, high frequency, low frequency, and exceptionality are interesting as well as important attributes of a synthesized association rule. In similar to Definitions 4 and 5, we propose type I and type II global exceptional association rules in multiple data sources as follows.

Definition 8 A high frequency association rule in multiple databases is type I global exceptional if it has low support. •

Definition 9 A heavy association rule in multiple databases is type II global exceptional if it has low frequency. •

Adhikari et al. (2011) have proposed iceberg pattern in multiple time-stamped databases. The authors have proposed an iceberg as a generalized notch that satisfies the following conditions: (1) The height of the generalized notch is greater than or equal to α, and (2) The width of the generalized notch is greater than or equal to β. Both α and β are user-defined thresholds. Authors have proposed an algorithm for mining icebergs in multiple time-stamped data sources.

7.3 Problem Statement

Let X be a type II global exceptional frequent itemset in D. Without any loss of generality, let X be extracted from $D_1, D_2, \ldots, D_k, 0 < k < n$. Support of X in D_i is $supp_a(X, D_i)$, $i = 1, 2, \ldots, k$. Then the average of these supports is obtained by the following formula:

$$avg\left(supp(X), D_1, D_2, \ldots, D_k\right) = \left(\sum_{i=1}^{k} supp_a(X, D_i)\right) \bigg/ k \qquad (7.3)$$

Database D_i is called an *exceptional source* with respect to the type II global exceptional frequent itemset X, if $supp_a(X, D_i) \geq avg(supp(X), D_1, D_2, \ldots, D_k)$, $i = 1, 2, \ldots, k$. We take an example to explain this issue. Let X be a global exceptional frequent itemset in D, and it has been extracted from $D_1, D_4,$ and D_7 out of ten data sources $D_1, D_2, \ldots, D_{10}$. Let $supp_a(X, D_1) = 0.09$, $supp_a(X, D_4) = 0.17$, and $supp_a(X, D_7) = 0.21$. Then, $avg(supp(X), D_1, D_4, D_7) = (0.09 + 0.17 + 0.21)/3 = 0.157$. The databases D_4 and D_7 are exceptional sources for type II global exceptional frequent itemset X, since $0.17 > 0.157$, and $0.21 > 0.157$. We state the problem as follows:

Let there be n data sources $D_1, D_2, \ldots, D_n$, and D be the union of these data sources. Let SFIS(D_i) be the set of frequent itemsets in D_i, $i = 1, 2, \ldots, n$. Find the type II global exceptional frequent itemsets in D using SFIS(D_i), $i = 1, 2, \ldots, n$. Also, report the exceptional sources for type II global exceptional frequent itemsetsi in D.

7.4 Related Work

Multi-database mining has been recently recognized as an important research topic in data mining (Zhang et al. 2004b; Adhikari et al. 2010a). Zhang et al. (2003) have proposed local pattern analysis for mining multiple large databases. Afterwards, we have proposed an extended model of local pattern analysis (Adhikari and Rao 2008c). The extended model provides a way to mine and analyze multiple large databases approximately. Later, we have proposed pipelined feedback technique (Adhikari et al. 2010a, b) that improves the accuracy of multi-database mining significantly.

Distributed data mining (DDM) algorithms deal with mining multiple databases distributed over different geographical regions. In the last few years, researchers have started addressing problems where the databases stored at different places cannot be moved to a central storage area for variety of reasons. In multi-database mining, there are no such restrictions. Distributed data mining environment often comes with different distributed sources of computation. The advent of ubiquitous computing (Greenfield 2006), sensor networks (Zhao and Guibas 2004), grid computing (Wilkinson 2009), and privacy-sensitive multiparty data (Kargupta et al. 2008) present examples where centralization of data is either not possible, or at least not always desirable.

In the context of support estimation of frequent itemsets, Jaroszewicz and Simovici (2002) have proposed a method for estimating supports of frequent itemsets using Bonferroni-type inequalities (Galambos and Simonelli 1996). Also, the maximum-entropy approach to support estimation of a general Boolean expression is proposed by Pavlov et al. (2000). However, these support estimation techniques are suitable for problems that deal with single database.

Existing parallel mining techniques could also be used to deal with multiple databases (Agrawal and Srikant 1994; Chattratichat et al. 1997; Cheung et al. 1996).

7.5 Synthesizing Support of an Itemset

Synthesizing support of an itemset in multiple databases is an important issue. First we present here the support estimation technique proposed in (Adhikari and Rao 2008c). In real databases, the trend of the customers' behaviour exhibited in one database is usually present in other databases. In particular, a frequent itemset in a database is usually present in some transactions of other databases even if it does not get reported there. The estimation procedure captures such trend and estimates the support of an itemset that fails to get reported in a database. The estimated support of a missing itemset usually reduces the error of synthesizing a frequent itemset in multiple data sources. If an itemset X fails to get reported from database D_1, then we assume that D_1 contributes some amount of support of X. The support of X in D_1 satisfies the following inequality: $0 \leq supp_a(X, D_1) < \alpha$. The procedure of finding an estimated support of an itemset, called *average low-support (als)*, is discussed below.

Let the itemset X be reported from m databases, $1 \leq m < n$. Without any loss of generality, we assume that X has been extracted from the first m databases. We shall use the average behaviour of the customers of the first m branches to estimate the average behaviour of the customers in remaining branches. Let $D_{1, m}$ be the union of $D_1, D_2, ..., $ and D_m. Then $supp_a(X, D_{1, m})$ could be viewed as the average behaviour of the customers of the first m branches with respect to X. Thus, $supp_a(X, D_{1, m})$ is computed according to the following formula.

$$supp_a(X, D_{1,m}) = \left(\sum_{i=1}^{m} supp_a(X, D_i) \times size(D_i) \right) \Big/ \sum_{i=1}^{m} size(D_i) \qquad (7.4)$$

One could estimate the support of X for each of the remaining databases as follows.

$$als(X, D_i) = \alpha \times supp_a(X, D_{1,m}), i = m+1, m+2, ..., n \qquad (7.5)$$

The technique discussed above might not be suitable for synthesizing type II global exceptional frequent itemsets. The reason is given as follows. A type II global exceptional frequent itemset X gets extracted from a few databases. During the process of synthesis, we need to estimate the supports of X for the remaining databases. So, the number of actual supports of X is much less than the number of estimated supports of X. Thus, the error of synthesizing the support of X in D might be high. Thus, we shall follow a different strategy for synthesizing support of an itemset in D. The strategy is explained as follows. We will mine databases at a reasonably low value of α. If the itemset X fails to get reported from D_i then we assume that $supp_a(X, D_i) = 0$, for some i. Here the itemset X is present in D_i, $i = 1, 2, ..., m$. Then the number of the transactions containing X in D_i is $supp_a(X, D_i) \times size(D_j)$, $i = 1, 2, ..., m$. We assume that the estimated number of the transactions containing X in D_i is 0, for $i = m + 1, m + 2, ..., n$. Thus, the estimated support of X in a database is given as follows:

$$supp_e(X, D_i) = \begin{cases} supp_a(X, D_i), i = 1, 2, ..., m \\ 0, i = m + 1, m + 2, ..., n \end{cases} \tag{7.6}$$

The synthesized support of X in D is determined with the use of the following formula.

$$supp_s(X, D) = \left(\sum_{i=1}^{n} supp_e(X, D_i) \times size(D_i) \right) \Big/ \sum_{i=1}^{n} size(D_i) \tag{7.7}$$

7.6 Synthesizing Type II Global Exceptional Itemsets

In this section, we present an algorithm for synthesizing type II global exceptional frequent itemsets in D. We discuss here various data structures required to implement the algorithm. Let N be the number of frequent itemsets in $D_1, D_2, ...,$ and D_n. The frequent itemsets are kept in a two dimensional array $SFIS$. The (i, j)-th element of $SFIS$ stores the j-th frequent itemset extracted from D_i, $j = 1, 2, ..., |SFIS(i)|$, $i = 1, 2, ..., n$. An itemset could be described by the following attributes: $itemset$, $supp$ and did. Here the attributes $itemset$, $supp$ and did represent itemset, support and database identification of the frequent itemset, respectively. Synthesized type II global exceptional frequent itemsets are kept in array $synFIS$. Each type II global exceptional itemset has been described by the following attributes: $itemset$, $ssupp$, $nSources$, $databases$, $nExSources$, and $exDbases$. The attributes $itemset$ and $ssupp$ represent the itemset and synthesized support of the type II global exceptional frequent itemset in D, respectively. The attributes $nSources$ and $databases$ store the number of sources of exceptional frequent itemsets and the list of identifications of source databases for a type II global exceptional frequent itemset, respectively. The attributes $nExSources$ and $exDbases$ store the number of exceptional sources and the list of identifications of exceptional sources for a type II global exceptional frequent itemset, respectively. The algorithm is presented below:

Algorithm 7.1   Synthesize type II global exceptional frequent itemsets in multiple databases.

procedure *Type-II-Exceptional-Frequent-Itemset-Synthesis (n, SFIS, μ_2, size, γ_1)*
Input:
n: number of data sources
SFIS: array of frequent itemsets
μ_2: threshold of high support
size: array of total number of transactions in different databases
γ_1: threshold of low extraction
Output:
Type II global exceptional frequent itemsets in *D*
01: collect all the frequent itemsets into array *FIS*;
02: sort frequent itemsets in *FIS* in non-decreasing order on itemset attribute;
03: calculate the total number of transactions into *totTrans*;
04: *nSynFIS* = 0; *curPos* = 1;
05: **while** (*curPos* ≤ *N*) **do**
06:     *i* = *curPos*; *count* = 0;
07:     *nTransCurFIS* = 0; *totSupp* = 0;
08:     **while** (*i* ≤ *curPos* + *n*) **do**
09:        **if** (*FIS(i).itemset* = *FIS(curPos).itemset*) **then**
10:           update *count, sources(count), totalSupp, nTransCurFIS, supports(count)*;
11:        **else** exit from the while-loop;
12:     **end if**
13:        increase *i* by 1;
14:     **end while**
15:     **if** ((*count / n*) < γ_1) **and** (*nTransCurFIS / totTrans* ≥ μ_2)) **then**
16:        increase *nSynFIS* by 1;
17:        update attributes *supp, itemset* and *nSources* of *synFIS(nSynFIS)*;
18:        *avgSupp* = *totalSupp / count*; *exCount* = 0;
19:        **for** *j* = 1 to *count* **do**
20:           *synFIS(nSynFIS).databases(j)* = *source(j)*;
21:           **if** (*supports(exCount)* ≥ *avgSupp*) **then**
22:              increase *exCount* by 1;
23:              *synFIS(nSynFIS).exDbases(exCount)* = *sources(j)* ;
24:           **end if**
25:        **end for**
26:        *synFIS(nSynFIS).nExSources* = *exCount*;
27:     **end if**
28:     *curPos* = *i*;
29: **end while**
30: sort itemsets in *synFIS*;
31: **for** *i* = 1 to *nSynFIS* **do**
32:     display details of *synFIS(i)*;
33: **end for**
34: **end procedure**

Let us explain the main processing flow of the above algorithm. The frequent itemsets with the same *itemset* attribute are kept consecutive in *FIS*. It helps processing one itemset at a time. A type II global exceptional frequent itemset is synthesized using the lines 5–29. The array *sources* is used to store the database identifications of all the databases that report the current frequent itemset. Also, the array *supports* is used to store the supports of the current frequent itemset in different databases. The information about the current itemset is obtained by the while-loop in lines 8–14. The information includes the number of extractions, database identifications of the source databases, supports in different databases, total supports, the number of transactions containing current frequent itemset in different databases. We explain the update-statement at line 10 as follows. The number of extraction of current frequent itemset, *count*, is increased by one. The database identification, *did*, of the current database is copied into cell *sources(count)*. Variable *nCurFIS* is added by expression $FIS(i).supp \times size$ $(FIS(i).did)$. Variable totalSupp is also added by expression $FIS(i).supp$. The support of frequent itemset in the current database is copied into *supports(count)*. We also explain the update-statement at line 17 as follows. The synthesized support of current type II global exceptional frequent itemset is obtained by expression *nCurFIS/totTrans*. The *itemset* attribute of current type II global exceptional frequent itemset is the same as the *itemset* attribute of current frequent itemset. The variable *count* is copied into *synFIS(nSynFIS).nSources*. The if-statement in lines 15–27 checks whether the current itemsets is a type II global exceptional frequent itemset in multiple data sources, and it synthesizes each type II global exceptional frequent itemset. All the frequent itemsets are processed using the lines 4–29. We sort global exceptional itemsets for better presentation at line 30. All the type II global exceptional itemsets are kept in non-decreasing order on the length of the itemset. Again, the itemsets of the same length are sorted on non-increasing order on support of the itemset. Finally, type II global exceptional itemsets and supports are displayed using lines 31–33. For every type II global exceptional frequent itemset, we display the data sources from which it has been extracted. Also, for every type II global exceptional frequent itemset, we also display the exceptional data sources from which it has been highly supported.

Theorem 7.1 *The time complexity of the procedure Type-II-Exceptional-Frequent-Itemset-Synthesis is maximum{$O(N \times log(N))$, $O(n \times N)$}, where N is the number of frequent itemsets extracted from n databases.*

Proof The time complexity of line 1 is $O(N)$, since there are N frequent itemsets in all the databases. Line 2 sorts N frequent itemsets in $O(N \times log(N))$ time. The time complexity of line 3 is $O(n)$, since there are n databases. The program segment lines 5–29 repeats maximum N times. Within this program segment, there is a while-loop and a for-loop. The while-loop in lines 8–14 takes $O(n)$ time. Also, the for-loop in lines 19–25 takes $O(n)$ time. Thus, the time complexity of the program segment lines 5–29 is $O(n \times N)$. Line 30 takes $O(N \times log(N))$ time for sorting maximum N synthesized type II global exceptional itemsets. To display the details of a type II global exceptional itemset it takes $O(n)$ time, since there are maximum

n sources of the itemset. Thus, the program segment in lines 31–33 take $O(n \times N)$ time. Therefore, the time complexity of the procedure *Type-II-Exceptional-Frequent-Itemset-Synthesis* is *maximum* $\{O(N \times log(N)), O(n \times N)\}$.•

In the following example, we execute the above algorithm step-by-step and demonstrate its functioning.

Example 7.1 Let D_1, D_2 and D_3 be three databases of sizes 4,000 transactions, 3,290 transactions, and 10,200 transactions respectively. Let *DB* be the union of D_1, D_2, and D_3. Assume that $\alpha = 0.1$, $\gamma = 0.4$, $\mu = 0.25$. The sets of frequent itemsets extracted from these databases are given as follows. $SFIS(D_1) = \{A(0.12), B(0.14), AB(0.11), C(0.20)\}$, $SFIS(D_2) = \{A(0.10), B(0.20), C(0.25), D(0.16), CD(0.12), E(0.16)\}$, $SFIS(D_3) = \{A(0.11), C(0.60), F(0.77)\}$. The symbol $X(\eta)$ is used to indicate that X is a frequent itemset in the local database with support η. We keep frequent itemsets in array *FIS* and sort them in terms of the *itemset* attribute. The sorted frequent itemsets are given in Table 7.1.

Here, *totTrans* is 17,490. We synthesize the frequent itemsets in *FIS* with the synthesized frequent itemsets given in Table 7.2.

In Algorithm 7.1, we maintain synthesized global exceptional frequent itemset in array *synFIS*. For the purpose of explanation, Table 7.2 has been introduced here. From Table 7.2, we find that the synthesized supports of itemsets C and F are high, since $supp_s(C, DB) \geq \mu$ and $supp_s(F, DB) \geq \mu$. Itemset F has been extracted from one out of three databases. Thus, F is low-voted, since $1/3 < \gamma$. Thus, the itemset F is a type II global exceptional frequent itemset in *DB*. •

In the following theorem, we determine time complexity of algorithm *IdentifyExPattern* (Zhang et al. 2004a) for comparing algorithm *IdentifyExPattern* with algorithm *Exceptional-FrequentItemset-Synthesis* theoretically.

Theorem 7.2 The algorithm *IdentifyExPattern* takes $O(n^2 \times N \times log(N))$ time, where N is the number of frequent itemsets extracted from n databases.

Proof We refer to algorithm *IdentifyExPattern*. Step 5 of the algorithm ranks candidate exceptional patterns based on their global supports. Step 4 calculates global support of a candidate exceptional pattern based on the number of databases that support the pattern. Step 1 counts the number databases that support a specific pattern. Thus, step 5 is based on step 4, and step 4 is based on step 1. Step 1 takes

Table 7.1 Sorted frequent itemsets in different databases

Itemset	A	A	A	B	B	C	C	C	D	E	F	AB	CD
Supp	0.12	0.10	0.11	0.14	0.20	0.20	0.25	0.60	0.16	0.16	0.77	0.11	0.12
Did	1	2	3	1	2	1	2	3	2	2	3	1	2

Table 7.2 Synthesized frequent itemsets in multi-databases

Itemset	A	B	C	D	E	F	AB	CD
Synthesized supp	0.11	0.07	0.44	0.03	0.03	0.45	0.03	0.02

$O(n)$ time for a specific pattern. This implies that step 4 takes $O(n \times n)$ time for each candidate exceptional pattern. Thus, step 5 takes $O(n^2 \times N \times log(N))$ time, and hence the theorem follows. •

From Theorems 7.1 and 7.2, we conclude that the proposed algorithm executes faster than algorithm *IdentifyExPattern*. We also compare these two algorithms experimentally in Sect. 7.8.

7.7 Error Calculation

To evaluate the proposed technique of synthesizing type II global exceptional frequent itemsets, we have measured amount of error occurred in an experiment. Error of the experiment is relative to the number of transactions (i.e., the size of the database), number of items, and length of a transaction in a database. Thus, the error of the experiment needs to be expressed along with the *ANT*, *ALT*, and *ANI* in a database, where *ANT*, *ALT*, and *ANI* denote the average number of transactions, average length of a transaction and average number of items in a database respectively. The error of the experiment is based on the global exceptional frequent itemsets in D. Let $\{X_1, X_2, ..., X_m\}$ be the set of type II global exceptional frequent itemsets in D. There are several ways one could define the error of an experiment. We have defined the following two types of errors.

1. Average Error (*AE*)

$$AE(D, \alpha, \mu, \gamma) = \frac{1}{m} \sum_{i=1}^{m} |supp_a(X_i, D) - supp_s(X_i, D)| \qquad (7.8)$$

2. Maximum Error (*ME*)

$$AE(D, \alpha, \mu, \gamma) = maximum\{ |supp_a(X_i, D) - supp_s(X_i, D)|, i = 1, 2, ..., m \} \qquad (7.9)$$

Actual support of X_i in D, $supp_a(X_i, D)$ is obtained by mining D using a traditional data mining technique, $i = 1, 2, ..., m$. Synthesized support of X_i in D, $supp_s(X_i, D)$ is obtained by the technique presented in Sect. 7.6, $i = 1, 2, ..., m$. Then we compute the error of synthesizing support of X_i in D as $| supp_a(X_i, D) - supp_s(X_i, D)|$, $i = 1, 2, ..., m$.

Example 7.2 With reference to Example 7.1, the itemset F is the only type II global exceptional frequent itemset present in D. Thus, $AE(D, 0.1, 0.25, 0.4) = ME(D, 0.1, 0.25, 0.4) = |supp_a(F, D) - supp_s(F, D)|$. •

7.8 Experiments

We have carried out several experiments to study the effectiveness of our approach. All the experiments have been implemented on a 2.8 GHz Pentium D dual processor with 512 MB of memory using visual C++ (version 6.0) software. The experimental results are presented on artificial and real datasets. We have constructed artificial database *check* to verify that the proposed algorithm works correctly. Each item is represented by an integer number to perform experiments more conveniently. Thus, a transaction in *check* is a collection of integer numbers separated by commas. The dataset *retail* (Frequent Itemset Mining Dataset Repository) is obtained from an anonymous Belgian retail supermarket store. Also, we have generated a synthetic dataset *SD100000*. We present some characteristics of these datasets in Table 7.3.

The symbols *NT*, *AFI*, *ALT* and *NI* denote the number of transactions, average frequency of an item, average length of a transaction and number of items in the dataset, respectively. Each of the above datasets has been divided into 10 databases for the purpose of carrying out experiments. These databases are called input databases. The proposed work is based on the frequent itemsets in the input databases. There are many algorithms (Agrawal and Srikant 1994; Han et al. 2000; Savasere et al. 1995) for mining frequent itemsets in a database. Thus, there exist many implementations (FIMI 2004) of mining frequent itemsets in a database.

The dataset *check* consists of 40 transactions. The input databases obtained from *check* are given as follows: $C_0 = \{\{1, 4, 9, 31\}, \{1, 4, 7, 10, 50\}, \{1, 4, 10, 20, 24\}, \{1, 4, 10, 23\}\}; C_1 = \{\{1, 4, 10, 34\}, \{1, 3, 44\}, \{1, 2, 3, 10, 20, 44\}, \{2, 3, 20, 39\}\}; C_2 = \{\{2, 3, 20, 44\}, \{2, 3, 45\}, \{2, 3, 44, 50\}, \{2, 3, 20, 44, 50\}\};$ $C_3 = \{\{3, 44\}, \{3, 19, 50\}, \{5, 7, 21\}, \{5, 8\}\}; C_4 = \{\{5, 41, 45\}, \{5, 49\}, \{5, 7, 21\}, \{5, 11, 21\}\}; C_5 = \{\{6, 41\}, \{6, 15, 19\}, \{11, 12, 13\}, \{11, 21, 49\}\};$ $C_6 = \{\{11, 19\}, \{21\}, \{21, 24, 39\}, \{22, 26, 38\}\}; C_7 = \{\{22, 30, 31\}, \{24, 35\}, \{25, 39, 49\}, \{26, 41, 46\}\}; C_8 = \{\{30, 32, 42\}, \{32, 49\}, \{41, 45, 59\}, \{42, 45\}\};$ $C_9 = \{\{42, 47\}, \{45, 49\}, \{47, 48, 49\}, \{49\}\}$. The input databases obtained from *retail* and *DS100000* are named as R_i and D_i, $i = 0, 1, ..., 9$. For the purpose of mining input databases, we have implemented apriori algorithm (Agrawal and Srikant 1994), since it is simple and popular. Some characteristics of these input databases (*DB*) are presented in Tables 7.4 and 7.5.

The global exceptional frequent itemsets corresponding to *check*, *retail* and *SD100000* are presented in Tables 7.6, 7.7 and 7.8 respectively.

Table 7.3 Dataset characteristics

Dataset	*NT*	*ALT*	*AFI*	*NI*
Check (*C*)	40	3.03	3.10	39
Retail (*R*)	88,162	11.31	99.71	10,000
SD100000 (*S*)	1,00,000	15.11	151.10	10,000

Table 7.4 The characteristics of databases obtained from *retail*

DB	NT	ALT	AFI	NI	DB	NT	ALT	AFI	NI
R_0	9,000	11.24	12.07	8,384	R_5	9,000	10.86	16.71	5,847
R_1	9,000	11.21	12.26	8,225	R_6	9,000	11.20	17.42	5,788
R_2	9,000	11.34	14.60	6,990	R_7	9,000	11.16	17.35	5,788
R_3	9,000	11.49	16.66	6,206	R_8	9,000	11.99	18.69	5,777
R_4	9,000	10.96	16.04	6,148	R_9	7,162	11.69	15.35	5,456

Table 7.5 The characteristics of databases obtained from *SD100000*

DB	NT	ALT	AFI	NI	DB	NT	ALT	AFI	NI
S_0	10,000	16.01	17.08	9,371	S_5	10,000	15.77	15.95	9,886
S_1	10,000	15.23	16.16	9,426	S_6	10,000	16.69	19.58	8,524
S_2	10,000	13.93	17.35	8,031	S_7	10,000	14.93	15.77	9,468
S_3	10,000	12.55	14.69	8,541	S_8	10,000	15.01	16.95	8,856
S_4	10,000	17.56	19.05	9,003	S_9	10,000	13.97	15.31	9,119

Table 7.6 Global exceptional frequent itemsets in $\{C_0, C_1, ..., C_9\}$ at $\alpha = 0.05$, $\gamma_1 = 0.4$ and $\mu_2 = 0.09$ (top 10)

Itemset	ssupp	Sources	Exceptional sources
{1}	0.175	C_0, C_1	C_0
{2}	0.175	C_1, C_2	C_2
{3}	0.200	C_1, C_2, C_3	C_1, C_2
{5}	0.150	C_3, C_4	C_4
{44}	0.150	C_1, C_2, C_3	C_1, C_2
{2,3}	0.150	C_1, C_2	C_2
{3,44}	0.150	C_1, C_2, C_3	C_2, C_3
{4}	0.125	C_0, C_1	C_0
{10}	0.125	C_0, C_1	C_0
{20}	0.125	C_0, C_1, C_2	C_1, C_2

Table 7.7 Global exceptional frequent itemsets in $\{R_0, R_1, ..., R_9\}$ at $\alpha = 0.02$, $\gamma_1 = 0.4$ and $\mu_2 = 0.1$

Itemset	ssupp	Sources	Exceptional sources
{2, 6, 9}	0.102	R_0, R_1	R_0
{2, 9, 41}	0.103	R_0, R_1, R_7	R_0, R_7
{6, 9, 41}	0.107	R_0, R_1, R_3	R_0, R_1
{8, 9, 271}	0.102	R_0	R_0
{9, 41, 48}	0.102	R_1	R_1

A global exceptional frequent itemset might not be supported with equal degree coming from the source databases. For example, the global exceptional frequent itemset {5} has been extracted from databases C_3 and C_4. But, the database C_4 reports itemset {5} to be exceptional to a very high extent.

Table 7.8 Global exceptional frequent itemsets in $\{S_0, S_1, \ldots, S_9\}$ at $\alpha = 0.02$, $\gamma_1 = 0.4$ and $\mu_2 = 0.05$ (top 10)

Itemset	*ssupp*	Sources	Exceptional sources
$\{22\}$	0.247	S_0, S_5, S_6	S_0
$\{32\}$	0.208	S_2, S_4, S_5	S_4, S_5
$\{45\}$	0.185	S_3, S_4, S_5	S_3
$\{4\}$	0.167	S_2, S_5	S_5
$\{49\}$	0.161	S_4, S_9	S_4
$\{22, 32\}$	0.158	S_0, S_2, S_5	S_0, S_2
$\{22, 45\}$	0.152	S_0, S_4, S_5	S_0, S_4
$\{22, 4\}$	0.132	S_0, S_2, S_5	S_0, S_5
$\{22, 49\}$	0.130	S_0, S_4	S_4
$\{32, 45\}$	0.121	S_2, S_4	S_2

We observe that some databases do no report any global exceptional frequent itemsets. On the contrary, some other databases are the source of many global exceptional frequent itemsets. In Tables 7.9, 7.10 and 7.11, we present the distributions of global exceptional frequent itemsets in $\{C_0, C_1, \ldots, C_9\}$, $\{R_0, R_1, \ldots, R_9\}$ and $\{S_0, S_1, \ldots, S_9\}$ respectively. In Table 7.9, we notice that three out of ten databases are not source of any global exceptional frequent itemset.

The distribution of global exceptional frequent itemsets in $\{R_0, R_1, \ldots, R_9\}$ is different from that in $\{C_0, C_1, \ldots, C_9\}$. In Table 7.10, we notice that seven out of ten databases are not the source of any exceptional global frequent itemset. In Table 7.11, we observe that the most of the databases are the source of some exceptional global frequent itemsets.

Table 7.9 Distribution of exceptional source for global exceptional frequent itemsets in $\{C_0, C_1, \ldots, C_9\}$ at $\alpha = 0.05$, $\gamma_1 = 0.4$ and $\mu_2 = 0.09$

Database	C_0	C_1	C_2	C_3	C_4	C_5	C_6	C_7	C_8	C_9
Number of sources of global exceptional frequent itemsets	7	8	10	4	2	1	1	0	0	0

Table 7.10 Distribution of global exceptional frequent itemsets in $\{R_0, R_1, \ldots, R_9\}$ at $\alpha = 0.02$, $\gamma_1 = 0.4$ and $\mu_2 = 0.1$

Database	R_0	R_1	R_2	R_3	R_4	R_5	R_6	R_7	R_8	R_9
Number of sources of global exceptional frequent itemsets	4	2	0	0	0	0	0	1	0	0

Table 7.11 Distribution of global exceptional frequent itemsets in $\{S_0, S_1, \ldots, S_9\}$ at $\alpha = 0.02$, $\gamma_1 = 0.4$ and $\mu_2 = 0.05$

Database	S_0	S_1	S_2	S_3	S_4	S_5	S_6	S_7	S_8	S_9
Number of sources of global exceptional frequent itemsets	11	0	6	9	1	5	7	3	3	5

Also, we have studied the number of global exceptional frequent itemsets in multi-databases at different γ_1 s. As we increase γ_1, we allow more frequent itemsets to be global exceptional. In Figs. 7.1, 7.2 and 7.3, we study the relationship between γ_1 and the number of global exceptional frequent itemsets in multiple databases.

We have observed that the number of global exceptional frequent itemsets do not vary much at different γ_1s. In fact, there is only one change in both the graphs of Figs. 7.1 and 7.2.

Also, we have studied the number of global exceptional frequent itemsets in multiple databases at different αs. We present experimental results in Figs. 7.4, 7.5 and 7.6. The number of global exceptional frequents in $\{C_0, C_1, \ldots, C_9\}$ remains fixed at 21 over different values of α.

But, we find a different trend with respect to the number of global exceptional frequent itemsets in $\{R_0, R_1, \ldots, R_9\}$. At lower and upper values of α, the number of global exceptional frequent itemsets is 0. Again, we get some global exceptional

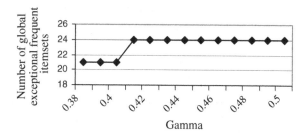

Fig. 7.1 Number of global exceptional frequent itemsets in $\{C_0, C_1, \ldots, C_9\}$ at $\alpha = 0.05$, and $\mu_2 = 0.1$

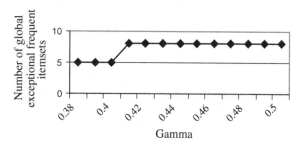

Fig. 7.2 Number of global exceptional frequent itemsets in $\{R_0, R_1, \ldots, R_9\}$ at $\alpha = 0.02$, and $\mu_2 = 0.1$

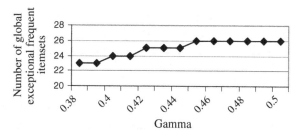

Fig. 7.3 Number of global exceptional frequent itemsets in $\{S_0, S_1, \ldots, S_9\}$ at $\alpha = 0.02$, and $\mu_2 = 0.05$

Fig. 7.4 Number of global exceptional frequent itemsets in $\{C_0, C_1, ..., C_9\}$ at $\gamma_1 = 0.4$, and $\mu_2 = 0.1$

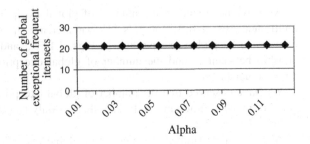

Fig. 7.5 Number of global exceptional frequent itemsets in $\{R_0, R_1, ..., R_9\}$ at $\gamma_1 = 0.4$, and $\mu_2 = 0.1$

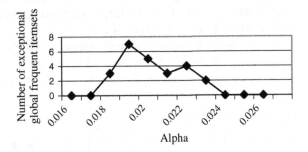

Fig. 7.6 Number of global exceptional frequent itemsets in $\{S_0, S_1, ..., S_9\}$ at $\gamma_1 = 0.4$, and $\mu_2 = 0.05$

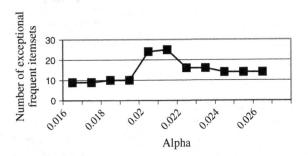

frequent itemsets for some middle values of α. Thus, there is no fixed relationship between the number of global exceptional frequent itemsets and the level of α.

Also, we have calculated the error of the experiments. In Table 7.12, we present the error of the experiments at a given value of tuple (α, γ, μ).

Table 7.12 Errors of the experiments at a given value of tuple (α, γ, μ)

Databases	α	γ_1	μ_2	(AE, *ANT, ALT, ANI*)	(ME, *ANT, ALT, ANI*)
$C_0, C_1, ..., C_9$	0.05	0.4	0.1	(0, 4, 3.025, 8.4)	(0, 4, 3.025, 8.4)
$R_0, R_1, ..., R_9$	0.02	0.4	0.1	(0.064, 8,816.2, 11.306, 5,882.1)	(0.085, 8,816.2, 11.306, 5,882.1)
$S_0, S_1, ..., S_9$	0.02	0.4	0.05	(0.052, 10,000.0, 15.11, 9,022.5)	(0.075, 10,000.0, 15.11, 9,022.5)

7.8.1 Comparison with the Existing Algorithm

In this section we complete a comparison between algorithms *IdentifyExPattern* (Zhang et al. 2004a) and *Exceptional-FrequentItemset-Synthesis*. We analyze and compare these two algorithms on the basis of experiments conducted with regard to the following two issues: (1) average error versus α, and (2) synthesizing time versus number of database.

7.8.1.1 Average Error

We have calculated AEs at different αs to study the relationship between them. Experimental results are presented in Figs. 7.7 and 7.8. We observe that there is no fixed relationship between AE and α.

In both the cases, algorithm *Exceptional-FrequentItemset-Synthesis* performs better than algorithm *IdentifyExPattern*. In dataset *check*, the global exceptional frequent itemsets are not uniformly distributed. The global exceptional frequent itemsets appear only in few databases, while they remain absent in the remaining databases. The proposed technique comes with an average error equal to 0 obtained

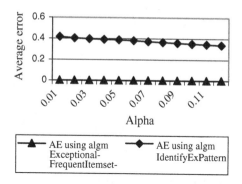

Fig. 7.7 Average error versus α for *check* at $\gamma_1 = 0.4$, and $\mu_2 = 0.1$

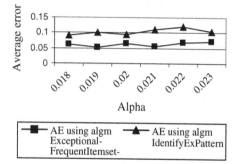

Fig. 7.8 Average error versus α for *retail* at $\gamma_1 = 0.4$, and $\mu_2 = 0.1$

for different αs, since the error of synthesizing each global exceptional frequent itemset in $\{C_0, C_1, ..., C_9\}$ is 0 (Fig. 7.8).

7.8.1.2 Synthesizing Time

Also, we have calculated the time for synthesizing global exceptional frequent itemsets by varying the number of databases. In Figs. 7.9, 7.10 and 7.11, we show time (in ms.) required for synthesizing global exceptional frequent itemsets in multiple databases. In case of the experiment conducted on $\{C_0, C_1, ..., C_9\}$, we observe that the synthesizing time does not increase as the number of databases increases. This is due to the fact that the size of each of the databases is very small. In fact, the time required for synthesizing global exceptional frequent itemsets in $\{C_0, C_1, ..., C_9\}$ is 0 ms., for both the algorithms.

Considering the results presented in Figs. 7.10 and 7.11, we could conclude that algorithm *Exceptional-FrequentItemset-Synthesis* executes faster than algorithm *IdentifyExPattern*. Also, it matches with the theoretical results. In general, the time for synthesizing global exceptional frequent itemsets either remains same or, increases as the number of databases increases.

Fig. 7.9 Synthesizing time versus number of databases obtained from *check* at $\gamma_1 = 0.4$, and $\mu_2 = 0.1$

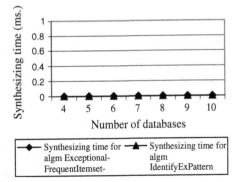

Fig. 7.10 Synthesizing time versus number of databases obtained from *retail* at $\alpha = 0.02$, $\gamma_1 = 0.4$, and $\mu_2 = 0.1$

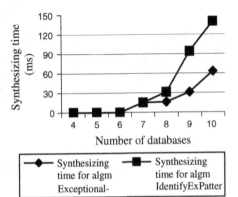

Fig. 7.11 Synthesizing time versus number of databases obtained from *SD10000I* at $\alpha = 0.02$, $\gamma_1 = 0.4$, and $\mu_2 = 0.05$

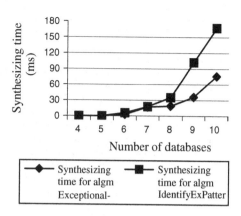

7.9 Conclusions

Synthesis of global exceptional patterns is an important component of a multi-database mining system. Many corporate decisions of a multi-branch company would depend on global exceptional patterns in branch databases. We have identified the short-comings of the existing concepts and the algorithm to identify global exceptional patterns. We have proposed a definition of a global exceptional frequent itemset. Also, we have introduced the notion of exceptional sources for a global exceptional frequent itemset. The proposed algorithm identifies global exceptional frequent itemsets and their exceptional sources in multiple databases. We have also compared proposed algorithm with the existing algorithm. The proposed algorithm performs better than the existing algorithm on the following issues: (1) error of the experiment and (2) execution time. We have observed that the proposed algorithm executes faster when the number of databases increases. Also, we have shown theoretically that the proposed algorithm executes faster than the existing algorithm. The proposed solution is simple and effective in synthesizing global exceptional frequent itemsets in multi-databases.

References

Adhikari A, Ramachandrarao P, Pedrycz W (2010a) Developing multi-database mining applications. Springer, London

Adhikari A, Rao PR (2008a) Mining conditional patterns in a database. Pattern Recogn Lett 29(10):1515–1523

Adhikari A, Rao PR (2008b) Efficient clustering of databases induced by local patterns. Decis Support Syst 44(4):925–943

Adhikari A, Rao PR (2008c) Synthesizing heavy association rules from different real data sources. Pattern Recogn Lett 29(1):59–71

Adhikari A, Rao PR, Prasad B, Adhikari J (2010b) Mining multiple large data sources. Int Arab J Inf Technol 7(3):241–249

Adhikari J, Rao PR, Pedrycz W (2011) Mining icebergs in time-stamped databases. In: Proceedings of the Indian international conferences on artificial intelligence, pp. 639–658

Adamo J-M (2001) Data mining for association rules and sequential patterns: sequential and parallel algorithms. Springer, Berlin

Agrawal R, Imielinski T, Swami A (1993) Mining association rules between sets of items in large databases. In: Proceedings of ACM SIGMOD conference on management of data, pp. 207–216

Agrawal R, Srikant R (1994) Fast algorithms for mining association rules. In: Proceedings of the international conference on very large data bases, pp. 487–499

Chattratichat J, Darlington J, Ghanem M, Guo Y, Hüning H, Köhler M, Sutiwaraphun J, To HW, Yang D (1997) Large scale data mining: Challenges, and responses. In: Proceedings of the third international conference on knowledge discovery and data mining, pp. 143–146

Cheung D, Ng V, Fu A, Fu Y (1996) Efficient mining of association rules in distributed databases. IEEE Trans Knowl Data Eng 8(6):911–922

FIMI (2004) http://fimi.cs.helsinki.fi/src/

Frequent Itemset Mining Dataset Repository. http://fimi.cs.helsinki.fi/data/

Galambos J, Simonelli I (1996) Bonferroni-type inequalities with applications. Springer, Berlin

Greenfield A (2006) Everyware: the dawning age of ubiquitous computing, 1st edition. New Riders Publishing, New Jersey

Han J, Pei J, Yiwen Y (2000) Mining frequent patterns without candidate generation. In: Proceedings ACM SIGMOD conference on management of data, pp. 1–12

Jaroszewicz S, Simovici DA (2002) Support approximations using Bonferroni-type inequalities. In: Proceedings of the sixth european conference on principles of data mining and knowledge discovery, pp. 212–223

Kargupta H, Han J, Yu PS, Motwani R, Kumar V (2008) Next generation of data mining. Springer, Berlin

Pavlov D, Mannila H, Smyth P (2000) Probabilistic models for query approximation with large sparse binary data sets. In: Proceedings of the sixteenth conference on uncertainty in artificial intelligence, pp. 465–472

Savasere A, Omiecinski E, Navathe S (1995) An efficient algorithm for mining association rules in large databases. In: Proceedings of the 21st international conference on very large data bases, pp. 432–443

Wang K, Zhou S, He Y (2001) Hierarchical classification of real life documents. In: Proceedings of the 1st (SIAM) international conference on data mining, pp. 1–16

Wilkinson B (2009) Grid computing: techniques and applications. CRC Press, Boca Raton

Wu X, Zhang S (2003) Synthesizing high-frequency rules from different data sources. IEEE Trans Knowl Data Eng 14(2):353–367

Wu X, Zhang C, Zhang S (2004) Efficient mining of both positive and negative association rules. ACM Trans Inf Syst 22(3):381–405

Wu X, Zhang C, Zhang S (2005a) Database classification for multi-database mining. Inf Syst 30(1):71–88

Wu X, Wu Y, Wang Y, Li Y (2005b) Privacy-aware market basket data set generation: a feasible approach for inverse frequent set mining. In: Proceedings of SIAM international conference on data mining, pp. 103–114

Zhang C, Zhang S (2002) Association rule mining: models and algorithms. Springer, Berlin

Zhang S, Wu X, Zhang C (2003) Multi-database mining. IEEE Comput Intel Bulletin 2(1):5–13

Zhang C, Liu M, Nie W, Zhang S (2004a) Identifying global exceptional patterns in multi-database mining. IEEE Comput Intel Bull 3(1):19–24

Zhang S, Zhang C, Wu X (2004b) Knowledge discovery in multiple databases. Springer, Berlin

Zhang S, Zhang C, Yu JX (2004c) An efficient strategy for mining exceptions in multi-databases. Inf Sci 165(1-2):1–20

Zhang S, Wu X (2011) Fundamentals of association rules in data mining and knowledge discovery. Wiley Interdisc Rew: Data Min Knowl Disc 1(2):97–116

Zhao F, Guibas L (2004) Wireless sensor networks: an information processing approach. Morgan Kaufmann, Los Altos

Chapter 8
Mining Icebergs in Different Time-Stamped Data Sources

Many organizations possess large databases collected over a long period of time. Analysis of such databases might be strategically important for further growth of the organizations. For instance, it might be of interest to learn about interesting changes in sales over time. In this chapter, we introduce a new pattern, called notch, of an item in time-stamped databases. Based on this notion, we propose a special kind of notch, called a generalized notch and subsequently, a specific type of generalized notch, called an iceberg, in time-stamped databases. We design an algorithm for mining interesting icebergs in time-stamped databases. We also present experimental results obtained for both synthetic and real-world databases.

8.1 Introduction

Many organizations collect transactional data continuously over a long period of time. A database grown over a long period of time might contain useful as well as interesting temporal patterns. By taking into account time aspect, many interesting applications/patterns such as surprising patterns (Keogh et al. 2002), discords (Keogh et al. 2006), calendar based patterns (Mahanta et al. 2008) have been recently reported. Surprising patterns, anomaly detection and discords could be considered as exceptional patterns occurring in a time series. These exceptional patterns form important as well as interesting contributions to temporal data mining. In this chapter, we study another kind of exceptional patterns in transactional time-stamped data. We define exceptional patterns and discuss how to mine them from time-stamped databases.

Though an analysis of time series data (Box et al. 2003; Brockwell and Richard 2002; Keogh 1997; Tsay 1989) has been intensively studied, the analysis of time-stamped data still calls for more research. Specifically, in the context of multiple time-stamped databases, little work has been reported so far. Therefore, there arises an urgent need to study multiple time-stamped databases. In Example 8.1, we observe an interesting type of temporal pattern in multiple time-stamped databases that needs to be fully analyzed.

A. Adhikari et al., *Data Analysis and Pattern Recognition in Multiple Databases*, Intelligent Systems Reference Library 61, DOI: 10.1007/978-3-319-03410-2_8, © Springer International Publishing Switzerland 2014

The support of an itemset (Agrawal et al. 1993) is defined as the fraction of transactions containing the itemset. It has been used extensively in identifying different types of patterns in a database. Some examples are association rule (Agrawal et al. 1993), negative association rule (Wu et al. 2004), and conditional pattern (Adhikari and Rao 2008). Nonetheless the support measure has a limited use in discovering some other types of patterns in a database. We illustrate this issue using the following example.

Example 8.1 Consider a company that maintains customers' transactions on a yearly basis. Many important problems can be studied given such yearly databases. Let item *A* be of our interest. In view of analyzing item *A* over the years, let us consider the sales series of *A* from the year 1970 to 1979:

(0.9, 1,000, 1970), (0.31, 2,700, 1971), (0.36, 2,500, 1972), (0.29, 3,450, 1973), (0.37, 1,689, 1974), (0.075, 7,098, 1975), (0.073, 8,900, 1976), (0.111, 6,429, 1977), (0.09, 9,083, 1978), (0.07, 10,050, 1979).

The individual components of each triple refer to the support of *A*, the number of transactions in the yearly database and the corresponding year, respectively.

The sales series of item *A* is depicted in Fig. 8.1. There is a significant downfall of sales from 1972 and rise in sales from the year 1975. Year 1975 is an important point (Pratt and Fink 2002) for the company. It is a significant down-to-up change in the sales series. It is not surprising to observe a significant up-to-down change in a sales series of an item. Such patterns in time-stamped series are interesting as well as important to investigate. They could reveal the sources of customers' purchase behavior that might provide an important knowledge to the organization.

At this point, one might be interested in knowing how a time series data differs from a time-stamped data. Transactional data are time-stamped data collected over time at no particular frequency (Leonard and Wolfe 2005). Whereas, time series data are time-stamped data collected over time at a particular frequency. For example, point of sales data could be considered as time-stamped data, but sales per month/ year could be considered as time series data. One could convert transactional data into time series data for the purpose of specific data analyses. The frequency associated with time series data varies from problem to problem. For future planning of business activities, one might need to analyze the past data of customer transactions collected over a long period of time. While analyzing the past data it is useful as well as important to figure out the abrupt changes in sales of an item along with time. Existing algorithms, as mentioned above, fail to detect these changes. Therefore, in this chapter our objective is to define such exceptional patterns and design an algorithm to extract such patterns from time-stamped databases.

Fig. 8.1 Sales of item *A* reported in consecutive years

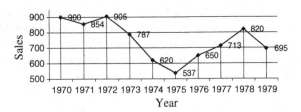

For the purpose of studying patterns in time-stamped databases one may need to handle multiple databases over time. One could call these time variant databases as *time databases*. In this context, the choice of time granularity is an important issue as the characteristics of temporal patterns is heavily dependent on this parameter. It is quite reasonable to consider the time granularity as one year, since a season re-appears on a yearly basis and the customers' purchase behaviour might vary from season to season.

Consider an established company having data collected over fifty consecutive years. The company might be interested in knowing the performance of different items over the years. Such analysis might help the company in devising the future strategies. Our objective is to identify abrupt changes in sales of each item (as defined in Sects. 8.4 and 8.5) over the years as depicted in Fig. 8.1.

The chapter is organized as follows. We discuss related work in Sect. 8.2. In Sect. 8.3, we introduce a new temporal pattern, called notch, of an item. Based on this pattern, we propose the concepts of generalized notch (Sect. 8.4) and iceberg notch (Sect. 8.5). We present another view of sales series in Sect. 8.6. In Sect. 8.7 we design an algorithm for mining icebergs in time-stamped databases. Experimental results are presented in Sect. 8.8.

8.2 Related Work

Temporal sequences appear in a vast range of domains ranging from engineering to medicine and finance, and the ability to model and extract information from them becomes crucial from a conceptual as well as applied perspective. Identifying exceptional patterns in time-stamped databases deserves much attention. In Sects. 8.4 and 8.5, we will propose two exceptional patterns in time-stamped databases.

There are mainly two general directions of temporal data mining (Roddick and Spillopoulou 1999). One concerns the discovery of casual relationships among temporally oriented events. Another one deals with the discovery of similar patterns within the same time sequence or among different time sequences. Sequences of events describe the behavior and actions of users or systems that can be collected in several domains. The proposed problem falls under the first category of the problems, since we are interested in identifying exceptional patterns by comparing sales in different years.

Agrawal et al. (1995) introduced the *shape definition language* (SDL), which used limited vocabulary such as {*Up, up, stable, zero, down, Down*} to describe different gradients in the series. The similarity of two time series is proportional to the length of the longest common sequence of words in their *SDL* representation. Such coarse information might not be always helpful. We define two exceptional patterns viz. a generalized notch and an iceberg notch.

Perng et al. (2000) proposed the landmark model where perceptually important points of a time series are used in its representation. The perceptual importance depends on the specific type of the time series. In general, sound choices for landmarks are local maxima and minima as well as inflection points. The advantage of using the landmark-based method is that this time representation is invariant to amplitude scaling and time warping. Some of the local maxima and minima might lead to higher level of exceptionality. Here we are concerned with defining similar exceptionalities in time-stamped databases.

There has been a significant amount of work on discovering temporal patterns of interest in sequence databases and time series databases. Temporal data mining is concerned with the analyses of data with an intention of finding patterns and regularities from a set of temporal data. In this context sequential association rule (Agrawal and Shrikant 1995), periodical association rule (Li and Deogun 2005), calendar association rule (Li et al. 2003) calendar-based periodic pattern (Mahanta et al. 2008) and up-to-date pattern (Hong et al. 2009) are some of the interesting temporal patterns reported in the recent time.

As noted in Sect. 8.1, support history of an item provides important information of an item over time. We have proposed an algorithm for clustering items in multiple databases based on their support history (Adhikari et al. 2009). We have introduced the notion of stability of an item based on its support history.

Lomet et al. (2008) integrated a temporal indexing technique, the TSB-tree, into Immortal DB to serve as the core access method. The TSB-tree provides high performance access and update for both current and historical data.

Keogh et al. (2006) proposed an algorithm for finding unusual time series where the notion of time discords is introduced. A time discord is a subsequence of a longer time series that is maximally different from all other subsequences of the series. Discords can be used to detect anomalies in an efficient way.

Many algorithms are designed incrementally to support time-dependent analysis of data. We have proposed algorithms incrementally to study an overall influence of a set of items on another set of itemsOverall influence of an item on another item in time databases (Adhikari and Rao 2010).

Castellana et al. (2007) proposed a new approach to performing change detection analyses based on a combination of supervised and unsupervised techniques is presented. Experimental results were reported for experiments carried out for image data. Wang et al. (2010) examined an unsupervised search method to discover motifs from multivariate time series data. The algorithm first scans the entire series to construct a list of candidate motifs in linear time. The list is then used to populate a sparse self-similarity matrix for further processing to generate the final selections. In contrast, the algorithm to be proposed is based on time-stamped data.

8.3 Notches in Sales Series

The change in sales of an item could be defined by the change of its support over time. The support of an item results in a *support history* (Bottcher et al. 2008) of the item in time databases. An analysis of a support history could be important in understanding customers' behavior (Adhikari et al. 2009). While dealing with the support history, the size of a database is an important issue. Support 0.129 from a database containing 1,000 transactions might be less important than the support 0.091 from a database containing 100,000 transactions. Thus, a mere analysis of the support history over time might not be effective in an application. One needs to analyze the supports along with the sizes in time databases. In Example 8.1, we observe that the support of A has been decreased from year 1970 to 1971. But the sales of A have been increased from the year 1970 to 1971. Hence, a negative change in support of an item might imply a positive change in frequency of the item. Thus, one needs to be careful in dealing with support history of an item in different databases.

Let us consider a company that has been operating for the last k years. For the purpose of studying temporal patterns, yearly databases could be constructed based on a time-stamp. Each of these databases corresponds to a specific period of time. Let D be the collection of customer transactions over k years. For the purpose of defining a temporal pattern we divide D into k yearly databases. Let DT_i be the database corresponding to the i-th year, $i = 1, 2, \ldots, k$. Each of these time databases is mined using a traditional data mining technique (Agrawal and Srikant 1994; Han et al. 2000). Mining time-stamped databases could help business leaders make better decisions by listening to their suppliers and/or customers after analyzing their transactions collected over time (Leonard and Wolfe 2005).

Over the years, an item may exhibit many consecutive data points having similar sales. As opposed to similar data patterns considering each data point, a limited yet meaningful number of points, may play a dominant role in many decision making problems. These meaningful data points could be defined in various ways, like average, peak, or slope of lines (Pratt and Fink 2002). In the context of the proposed problem, such compression of data points seems to be irrelevant. Given the sales series of an item, one might be interested in identifying abrupt changes in the sales series. The goal of our considerations is to define a new type of pattern based on abrupt variation of sales of an item over the years and to design an algorithm to mine such patterns in time databases.

Over the years, there may exist many ups and downs in sales of an item. One might be interested in identifying abrupt changes of sales in different years. It might be helpful to figure out the causes behind it and to take actions accordingly. Let $s_i(A)$ be the sales of item A for the i-th year, $i = 1, 2, \ldots, k$. We define the change in sales series of item A at year i as follows (Singh and Stuart 1998).

The change in sales series at year i is *increasing* if

$$s_{i-1}(A) < s_i(A) < s_{i+1}(A), \quad i = 2, 3, \ldots, k-1. \tag{8.1}$$

The change of sales series at year i is *decreasing* if

$$s_{i-1}(A) > s_i(A) > s_{i+1}(A), \quad i = 2, 3, \ldots, k-1. \tag{8.2}$$

The change of sales series at year i is *altering* if

$$s_{i-1}(A) < s_i(A) \quad \text{and} \quad s_i(A) > s_{i+1}(A), \tag{8.3}$$

or,

$$s_{i-1}(A) > s_i(A) \quad \text{and} \quad s_i(A) < s_{i+1}(A), \quad i = 2, 3, \ldots, k-1. \qquad (8.4)$$

The notion of *strict extrema* (Fink and Gandhi 2007) at a year corresponding to an item is defined as follows.

Let $s_i(A)$ be the amount of sales of an item A at year i, $i = 1, 2, \ldots, k$. There exists a *strict extrema* at year i for the item A if the change of support history of A at year i is altering. Based on the concept of strict extrema, we define a notch as follows.

Definition 8.1 There exists a *notch* at year i for the item A if there is a strict extrema at year i in the sales series of item A. •

Let $\Delta s_i(A)$ be the difference in sales of item A between the years i and $i - 1$. Now we propose a few definitions as follows.

Definition 8.2 Let there exists a notch at year i for item A. The notch at year i for item A is *downward* if $\Delta s_i(A) < 0$ and $\Delta s_{i+1}(A) > 0$. •

Definition 8.3 Let there exists a notch at year i for item A. The notch at year i for item A is *upward* if $\Delta s_i(A) > 0$ and $\Delta s_{i+1}(A) < 0$. •

Itemset (Agrawal et al. 1993) could be considered of as a basic type of pattern present in a transactional database. Many important as well as interesting patterns are based on itemset patterns. Similarly an upward/downward notch could be considered as a basic type of pattern in time databases. In Sect. 8.4, we illustrate how the notion of notch could help analyzing a special type of trend in time databases. Thus, it becomes important to mine notches in time databases.

8.3.1 Identifying Notches

Notches in sales series of an item could be considered as basic building blocks to construct temporal patterns. Later we introduce the notion of two interesting temporal patterns viz., generalized notches and iceberg notches in time databases. One could simply scan the sales series of an item to identify its interesting notches. Let n and k be the number of items in time databases and the number of time-stamped (yearly) databases, respectively. Then the time complexity of identifying notches is $O(n \times k)$.

8.4 Generalized Notch

Based on the concept of notch, we present here the notion of a generalized notch in time databases. Let us refer to Fig. 8.1. There are two downward notches in the years 1971 and 1975 having sales 854 and 537, respectively. The concept of notch can be generalized based on strict extrema as clarified in Sect. 8.3. One could notice in Fig. 8.1 that the downward notch in the year 1975 is wider than that of 1971. The width of a downward generalized notch is based on the two consecutive local maxima within which the downward generalized notch is enclosed. The width of the downward generalized notch in 1975 is $1978 - 1972 = 6$. Similarly, the width of an upward generalized notch is based on the two consecutive local minimums within which the upward generalized notch is enclosed. The width of the upward generalized notch in 1972 is $1975 - 1971 = 4$. Based on the above discussion, we define a width of a generalized notch as follows (Adhikari et al. 2011).

Definition 8.4 Let there be a generalized downward (upward) notch in the year i. Also, let the generalized downward (upward) notch be enclosed with the local maximums (minimums) in the years i_1 and i_2 ($>i_1$). The width of the generalized notch in the year i is equal to $i_2 - i_1$. •

The width of a generalized notch could be divided into left width and right width. The left width and right width are equal to ($i - i_1$) and ($i_2 - i$), respectively. In case of downward generalized notch the sales value gradually decreases, and then attains the minimum value, and then it gradually increases. Thus, the change of sales value between two consecutive years seems to be an important characteristic of a generalized notch. In this regard, one might be interested in the change of sales for a year as compared to its previous year. Also, the sales at year i as compare to sales of year i_1 and i_2 are important characteristics of a generalized notch. Accordingly, one could define left-height and right-height of the generalized notch in the year i for an item A as follow: *left-height* $(A, i) = |sales(A, i_1) - sales(A, i)|$, and *right-height* $(A, i) = |sales(A, i) - sales(A, i_2)|$. We define the height of a generalized notch as follows (Adhikari et al. 2011).

Definition 8.5 Let there exists a generalized downward (upward) notch in the year i. Also, let the generalized downward (upward) notch be enclosed with the local maximums (minimums) i_1 and i_2 ($>i_1$). The height of the generalized notch in the year i is equal to *maximum* {*left-height* (A, i), *right-height* (A, i)}. •

Based on the concept of generalized notch, we focus on the notion of iceberg.

8.5 Iceberg Notch

An analysis of sales series of items is an important issue. In view of performing this task, one could analyze the sales series for each item. In analyzing a sales series in-depth, it is evident that an existence of a notch might be an indication of a bigger notch. This represents an exceptionality of sales of an item. Based on such an exception, we define iceberg in time databases as follows (Adhikari et al. 2011).

Definition 8.6 An *iceberg* notch is a generalized notch that satisfies the following conditions: (1) The height of the generalized notch is greater than or equal to α, and (2) The width of the generalized notch is greater than or equal to β. Both α and β are user-defined thresholds. •

An iceberg notch is a generalized notch having a larger width and a lager height. The concept of iceberg in data management is not new. For example, iceberg queries (Han and Kamber 2001) are commonly used in data mining, particularly in market basket analysis.

Let us illustrate the concept of an iceberg using an example. Let the value of α and β be set to 300 and 5, respectively. Also, let the values of i, i_1 and i_2 be 1975, 1972, and 1978, respectively (with respect to Fig. 8.1). We observe that *left-Height*$(A, 1975) = |sales(A, 1972) - sales(A, 1975)| = |905 - 537| = 368$ and *rightHeight*$(A, 1975) = |sales(A, 1975) - sales(A, 1978)| = |537 - 820| = 283$, respectively. Therefore, the height of the iceberg is *maximum* $\{368, 283\}$ i.e. $368 \geq \alpha$. Also, $i_2 - i_1 = (1978 - 1972) = 6 \geq \beta$. So, there exists an interesting downward iceberg notch in the year 1975. We also observe an upward notch in the year 1972 with height $368 \geq \alpha$ and width $(1975 - 1971) = 4 < \beta$. So, the upward notch in the year 1972 is not an iceberg. We state the problem as follows.

Let there are k time databases. Let I be the set of all items in time databases. There exists a sales series of each item in I, consisting of k sales values, one for each time database. Mine interesting iceberg notches in sales series for each item in I.

8.6 Sales Series

A sales series of an item might provide interesting information about the item. It is basically the same as the support history of the item. As noted above, in many problems, it is preferable to analyze sales series rather than looking at the support history of an item. Many temporal patterns might originate by analyzing such types of temporal series.

8.6.1 Another View at Sales Series

Each data in a sales series can be mapped onto a member in the set $\{-1, 0, 1\}$ by comparing the data with the previous data in the same series. Thus, a time-stamped series could be mapped into a ternary series. This provides a simplified view of the original time-stamped series data. Such simplified view might provide some useful information. The procedure for mapping a time-stamped series into a ternary series is illustrated in the following example.

Example 8.2 Consider the sales data given in Example 8.1. The sales of item A in 1971 decreased from the sales in 1970. We note this behavior by putting -1 in the ternary series of item A corresponding to year 1971. The sales of item A in 1972 increased over the sales in 1971. We note this behaviour by putting $+1$ in the ternary series of item A corresponding to year 1972. If the sales of item A in any year remains same as that of previous year then we note this behaviour by putting 0 in the ternary series. Thus, we obtain the ternary series (TS) of item A in the following form:

Year	1970	1971	1972	1973	1974	1975	1976	1977	1978	1979
$TS(A)$		-1	$+1$	-1	-1	-1	$+1$	$+1$	$+1$	-1

In the above series, one can observe the existence of a generalized notch. The width of a downward generalized notch can be obtained from a run of -1' s and the subsequent run of $+1$' s. The width is obtained by adding the number of -1' s in the first run and the number of $+1$' s in the second run. A similar procedure can be followed for finding width of an upward generalized notch. In $TS(A)$ we observe a downward generalized notch in 1975. The width of this notch is equal to $3 + 3 = 6$. Also, there exists an upward generalized notch in 1978 having width of 4.

A slightly different procedure could also be followed for obtaining a ternary series corresponding to a sales series of an item. Let x and y be the sales for the year 1970 and 1971, respectively. Let δ be the level of significance of difference in sales. We put $+1$ in the ternary series of the item corresponding to year 1971, if $y - x > \delta$. We put -1 in the ternary series of the item corresponding to year 1971, if $x - y > \delta$. We put 0 in the ternary series of the item corresponding to year 1971, if $|x - y| \leq \delta$. The method of obtaining a ternary series using this procedure might be useful in many situations, since a small change in sales value might be insignificant in many situations. This procedure seems to be more realistic than the previous one.

8.6.2   Other Applications of Icebergs

We have mined icebergs from binary transactional datasets. In binary transactions the database can be viewed as records with Boolean attributes, where each attribute corresponds to a single item. Further, in the record of a transaction, the attribute corresponding to an item is true if and only if the transaction contains the item, otherwise false. But this type of datasets has got limited usage, since in a real life the items are often purchased multiple times. Therefore, in reality most of the market basket data are non-binary. Nowadays researchers from the data mining community are more concerned with qualitative aspect of attributes (e.g. significance, utility) as compared to considering only quantitative ones (e.g. number of appearances in a database etc.), because qualitative properties are required in order to fully exploit the attributes present in the dataset. By using this measure, icebergs can also be discovered. This utility measure calculates the actual profit value of an item in a non-binary transaction database. Thus, the same approach could be followed to mine the icebergs from the total utility value of the itemsets.

Icebergs could also be mined from other type of datasets such as rainfall data and crops production. To mine icebergs from rainfall data, yearly total rainfall is accumulated for a particular region. Similarly, total amount of particular crops are summed up for a specific crop in a region.

8.7 Mining Icebergs in Time-Stamped Databases

Let there are n items in time databases. For each item in time databases there exists a time-stamped series containing k data. In this section we are interested in identifying icebergs in each time-stamped series. For mining icebergs in time databases, we make use of an existing frequent itemset mining algorithm (Agrawal and Srikant 1994; Han et al. 2000). For the requirement of proposed problem one needs to mine the frequencies of each item in the time databases.

8.7.1 Non-incremental Approach

Based on the discussion held in the previous sections, we design an algorithm for mining icebergs in time databases. Let n and k be the number of items in time databases and the number of time databases, respectively. In this algorithm, we use a two-dimensional array F for storing frequencies of all items in time databases. F consists of n rows and $k + 1$ columns. The first column contains the items in time databases. For example, $F(i)(1)$ contains the i-th item in time databases, $i = 1, 2, ..., n$. The i-th row of F contains i-th item and its frequencies in k time databases. For example, $F(i)(j)$ contains the frequency of i-th item in $(j - 1)$-th database, $j = 2, 3, ..., k + 1$. Therefore, we need to check the existence of a generalized notch using the values in the columns from 2 to $k + 1$.

For the purpose of computing interestingness of a generalized notch, we determine the change of sales of a local minimum (maximum) with respect to its previous and next local maximum (minimum). During the process of mining icebergs, the generalized notches are kept in array GN. A generalized notch can be described by the following attributes: left year (*leftYear*), right year (*rightYear*), item (*item*), year of occurrence (*year*), type of generalized notch (*type*), change of sales at the year of occurrence with respect to the previous local extremum (*leftHeight*), change of sales at the year of occurrence with respect to the next local extremum (*rightHeight*), width of generalized notch (*width*) and the sales at the year of occurrence (*sales*). The goal of the proposed algorithm is to find all the interesting icebergs for each item in time databases. The algorithm is given as follows (Adhikari et al. 2011).

Algorithm 8.1 Mine icebergs in time-stamped databases.
procedure MineIcebergs (k, F, α, β)
  Inputs: k, F, α, β
  k: number of yearly databases
  F: array of frequencies of items in yearly databases
  α: user-defined threshold of height of a generalized notch
  β: user-defined threshold of width of a generalized notch
  Outputs:
  Interesting icebergs in time databases
  01: **let** *index* = 1;
  02: **for** $i = 1$ to n **do**
  03:   **let** $j = 2$; **let** *left* = 2; **let** *flat* = false; **let** *prevDown* = false; **let** *prevUp* = false;
  04:   **while not** end of the sales series corresponding to i-th item **do**
  05:     **if** there is a downward trend **and** *prevUp* is false **then**
  06:       find *mid*, *leftWidth*; **let** *prevDown* = true;
  07:       compute *leftHeight*; **go to** 04;
  08:     **end if** {05}
  09:     **if** there is an upward trend **and** *prevDown* is true **then**
  10:       find *left*, *mid*, *right*, *leftWidth*, *rightWidth*; **let** *prevDown* = false;
  11:       **let** *leftHeight* = *rightHeight*;
  12:       compute *rightHeight*;
  13:       *GN(index).type* = *down*; **go to** 30;
  14:     **end if** {09}
  15:     **if** there is an upward trend **and** *prevDown* is false **then**
  16:       **let** *prevUp* = true; find *mid*, *leftWidth*;
  17:       compute *leftHeight*; **go to** 04;
  18:     **end if** {15}
  19:     **if** there is a downward trend **and** *prevUp* is true **then**
  20:       find *left*, *mid*, *right*, *leftWidth*, *rightWidth*;
  21:       **let** *prevUp* = false; **let** *prevDown* = true;
  22:       **let** *leftHeight* = *rightHeight*;
  23:       compute *rightHeight*;
  24:       *GN(index).type* = *up*; **go to** 30;
  25:     **end if** {19}
  26:     **if** the sales of the j-th and $(j+1)$-th year remain same **then**
  27:       find *left*; **let** *flat* = true; **let** *prevDown* = false; **let** *prevUp* = false;
  28:       **go to** 04;
  29:     **end if**
  30:     **if** *flat* is false **then**
  31:       compute *height* as defined in Definition 5;
  32:       **if** the current generalized notch satisfies the criteria α and β **then**
  33:         store it in *GN(index)*; increase *index* by 1;
  34:       **end if**
  35:       **if** the current generalized notch is downward **then**
  36:         **let** *prevDown* = false; **let** *prevUp* = true;
  37:       **else if** the current generalized notch is upward **then**
  38:         **let** *prevDown* = true; **let** *prevUp* = false;
  39:       **end if**
  40:       **end if** {35}
  41:     **end if** {30}
  42:   **end while** {04}
  43: **end for** {02}
  44: display icebergs from *GN*;
end procedure

The lines 2–43 are repeated for each item in time databases. In each repetition, the interesting icebergs corresponding to an item are identified. The variable *index* is used to index array *GN*. The variable *j* is used to keep track of current sales data of the item under consideration. The starting value of *j* is 2, since the sales data for the first year of an item is kept starting from column number 2 of array *F*. We use three Boolean variables viz., *flat*, *prevUp* and *prevDown*. While identifying downward (or, upward) generalized notches, we first go through its left leg of a generalized notch. After reaching its minimum/maximum value, if the next point also attains the same value then *flat* becomes true. The width of a generalized notch is determined by the following years: left year (*left*), middle year (*mid*), and right year (*right*). Accordingly, the width of a generalized notch has two components: left width (*leftWidth*) and right width (*rightWidth*). After identifying the left leg of a downward (or, upward) generalized notch, *prevDown* (or, *prevUp*) becomes true. After storing the details of the current generalized notch *index* gets increased by 1 (line 33). We identify generalized notches for each item in time databases. For this purpose, we introduce a for-loop at line 02 which ends at line 43. Some lines, e.g. lines 40–43, are ended with a number enclosed in curly brackets, to mark the ends of composite statement starting with the line number kept in curly bracket.

Lines 2 and 4 repeat for *n* and *k* times respectively. In other words, the sales series corresponding to each item is processed for identifying icebergs. Thus, the time complexity of lines 1–43 is $O(n \times k)$. Again, the time complexity of line 44 can not be more than $O(n \times k)$, since the number of interesting icebergs is always less than $n \times k$.

Theorem 8.1 *Correctness of the MineIcebergs algorithm.*

Proof Consider that there are *n* items in *k* time-stamped databases. Each sales series is processed using lines 2–43. For the purpose of mining interesting icebergs, each sales series is checked completely by applying a while-loop shown in lines 4–42. A sales series can start with one of the following three ways: (a) showing a downward trend, (b) showing an upward trend (c) remained at a constant level. The algorithm handles each of these cases separately.

Case (a) has been checked at the line numbered as 05. Once a downward trend changes we again go back to while-loop at line 04 for finding one of the following two possibilities: an upward trend and a constant sales.

Case (b) has been checked at the line number 15. Once the upward trend changes we again go back to while-loop at line 04 for finding one of the following two possibilities: a downward trend and a constant sales.

Case (c) has been checked at the line number 26. Once the flatness changes we again go back to while-loop at line 04 for finding one of the following two possibilities: a downward trend and an upward trend.

Once the left leg of a downward generalized notch is detected in lines 5–8, its right leg is detected in lines 9–14. When the left leg of an upward generalized notch is detected in lines 15–18, its right leg is detected in lines 19–25. After detecting a generalized notch at lines 13 and 24, we go to line 30 for detecting its interestingness and re-initializing required Boolean variables.

Thus, the above algorithm considers all the possibilities that would arise in each sales series.

8.7.2 Incremental Approach

An incremental approach seems to be a natural way of designing an algorithm for mining icebergs in time-stamped databases. Every year we need to mine a yearly database, and the frequencies of all the items are computed. As a result, one more term in the frequency series of each item gets added. Thus, one needs to mine only the current database for the purpose of making an analysis based on the entire database. As a result one can mine icebergs using a cost-effective and faster algorithm.

Let ΔD be the database for the current year. In this algorithm, we use a two-dimensional array ΔF for storing frequencies of all items in the current yearly database. It consists of n rows and 2 columns. The first column contains the items in the current database and the second column contains the frequencies of the items. Similar to Algorithm 1, array F contains the frequencies of items of previous databases. Array ΔF is merged with the array F, and the frequency are stored in $(k + 2)$-th column. It can be considered as a preprocessing task of the proposed incremental mining algorithm. Currently, F consists of n rows and $k + 2$ columns.

For the purpose of computing interesting icebergs incrementally, we use array GN, where all the interesting icebergs for all items are stored for k years. The attributes of a generalized notch have already been described in the previous section. The goal of the proposed algorithm is to find all the interesting icebergs for each item in time databases incrementally. We have made use of procedure *MineIcebergs* while designing an incremental algorithm for mining icebergs. The incremental algorithm is given below.

Algorithm 8.2 Mine icebergs incrementally in time-stamped databases.
procedure MineIcebergsIncrementally (F, ΔF, GN, α, β)

Inputs: F, ΔF, GN, α, β

F: array of frequencies of items in the previous yearly databases

ΔF: array of frequencies of items in the current yearly database

GN: array of in interesting icebergs in previous yearly database

α: user-defined threshold of height of a generalized notch

β: user-defined threshold of width of a generalized notch

Outputs:

Interesting icebergs in time databases

```
01:  copy frequencies of items in ΔF into (k+2)-th column of F
02:  for j = 1 to n do
03:    if there exists an iceberg of j-th item in GN then
04:       go to the last iceberg I of j-th item in GN;
05:       if rightYear of I is equal to (k+1) then
06:         if (I is upward) and (sales of (k+2)-th year is less than rightYear) then
07:           update current iceberg;
08:         else call MineIcebergs (k+3-year, F(year, k+2), α, β);
09:         end if
10:         if (I is downward) and (sales of (k+2)-th year is greater than rightYear) then
11:           update current iceberg;
12:         else call MineIcebergs (k+3-year, F(year, k+2), α, β);
13:         end if
14:       else
15:         go to the last iceberg I of j-th item in GN;
16:         call MineIcebergs (k+3-year, F(year, k+2), α, β);
17:       end if {05}
18:    else call MineIcebergs (k+1, F(2, k+2), α, β);
19:    end if {03}
20:  end for {02}
21:  display icebergs from GN;
     end procedure
```

Line 1 merges ΔF with F so that F now contains n rows and $k + 2$ columns. Line numbers 2–20 repeats for each item present in the database. Lines 3–4 are used to check whether an iceberg of the current item is present in GN, and go to the last iceberg if it has been found. Lines 5–13 are used to update the last iceberg of current item with the help of current data kept in $(k + 2)$-th column. If the last iceberg of the current item occurs in the middle, then we go to the last iceberg and then call *MineIcebergs* (lines 15–16). In case if there is no iceberg for the current item then we call *MineIcebergs* using all the data kept in columns 2 to $(k + 2)$. Lines 2–20 execute for n times. Among these statements, line 18 has the maximum time complexity i.e., $O(n \times k)$. Therefore, the complexity of *MineIcebergsIncrementally* is $O(n^2 k)$.

The incremental algorithm appears to be more time consuming than the initial algorithm *MineIcebergs*. Both the algorithms are executed in the main memory and do not require any secondary memory access. These algorithms are required to follow a data mining algorithm that actually reads data from the secondary storage. Before applying algorithm *MineIcebergsIncrementally*, only one database i.e. the database for the $(k + 1)$-th year needs to be mined. But if we apply algorithm *MineIcebergs* then all the previous databases i.e. the databases for the first year to $(k + 1)$-th year need to be mined. Thus, the algorithm *MineIcebergsIncrementally* practically saves a significant amount of time.

8.8 Significant Year

In the scenario of possessing time-stamped data over several years, the organization might be interested in knowing the years when sales of a large number of items either increase or decrease as well as both increase and decrease. These years have importance in the history of the organization. Further analysis of such rare events might influence the strategies of the organization. In view of this, we define a significant year as follows.

Definition 8.7 Let there be n items transacted over the years. Year i is significant if the ratio of the number of icebergs and n exceeds user-defined threshold δ. •

δ is a fraction lies between 0 and 1. Such interestingness condition could also be stated with respect to either upward iceberg or downward iceberg.

For the purpose of mining significant years, we use array GN, where all the interesting icebergs are stored. We use the attributes *item*, *year*, and *type* of GN to mine significant years. A significant year can be described by the following attributes: *year* (year), *type* (upward/downward icebergs). Variable *count* is used to store total number of (upward/downward icebergs) in a particular year. Array *SigYear* is used to store significant years. The variables i, j, and *index* have been used to access arrays F, *SigYear* and GN, respectively. The proposed algorithm finds all the significant years that are attributed by large number of icebergs. The algorithm is presented below.

Algorithm 8.3 Mine the significant years.
procedure MineSignificantYears (n, k, δ, F, GN)
    Inputs: GN
    n: total number of items in the database
    k: number of yearly databases
    δ: user-defined threshold
    F: array of frequencies of items in the yearly databases
    GN: array of interesting icebergs
    Outputs:
    Significant years in time databases

```
01: let j = 1;
02: for i = 2 to (k+1) do
03:    let index = 1; let count = 0;
04:    while not end of GN do
05:      if F(1, i) = GN(index).year then
06:        increase count by 1;
07:      end if
08:      increase index by 1;
09:    end while
10:    if (count / n ≥ δ) then
11:      SigYear(j).year = F(1, i); SigYear(j).count = count; increase j by 1;
12:    end if
13: end for
14: display years from SigYear;
end procedure
```

Algorithm *MineSignificantYears* finds significant years with respect to both upward and downward icebergs. One might be interested in finding significant years with respect to upward or downward iceberg. The algorithms finds a number of icebergs year-wise; see lines 5–8. Then the interestingness of the year is checked at line 10. Let the number of icebergs in *GN* be *p*. Then the complexity of the algorithm is $O(k \times p)$.

8.9 Experimental Studies

We have carried out several experiments for mining generalized notches in different databases. All the experiments have been implemented on a 1.6 GHz Pentium IV with 256 MB of memory using visual C++ (version 6.0) software. We present experimental results using four real databases and two synthetic databases. The databases *mushroom, retail* (Frequent itemset mining dataset repository) *ecoli* and *BMS-WebView-1* are real-world databases. The real databases *BMS-Web-View-1* can be found from KDD CUP 2000 (Frequent itemset mining dataset repository). Database *ecoli* is a subset of *ecoli database* (UCI ML repository). The synthetic dataset *T10I4D100K* was generated using the generator from the IBM Almaden Quest research group. *Random-68* is also a synthetic database and has been generated for the purpose of conducting experiments. The characteristics of these databases are given in Table 8.1.

The symbols used in Tables 8.1 and 8.2 come with the following meaning: *D, NT, ALT, AFI,* and *NI* denote database, the number of transactions, average length of a transaction, average frequency of an item, and number of items, respectively. The databases *mushroom, ecoli, random-68* and *retail* have been divided into 10 sub-databases, called yearly databases, for the purpose of conducting experiments. The databases *BMS-WebView-1* and *T10I4D100K* have been divided into 20 databases. The databases obtained from *mushroom, ecoli, random-68* and *retail* are named as M_i, E_i, R_i, and Rt_i, $i = 0, 1, ..., 9$. The databases obtained from *BMS-WebView-1* and *T10I4D100K* are named as B_i and T_i, $i = 0, 1, ..., 19$. We present some characteristics of the input databases in Table 8.2.

In Tables 8.3, 8.4, 8.5 and 8.6 we have represented upward generalized notch as '*u*' and downward generalized notch as '*d*'. In *mushroom*, there are many items

Table 8.1 Database characteristics

Database	NT	ALT	AFI	NI
mushroom (M)	8,124	24.000	1,624.800	120
ecoli (E)	336	7.000	25.835	91
random-68 (R)	3,000	5.460	280.985	68
retail (Rt)	88,162	11.306	99.674	10,000
BMS-WebView-1 (B)	149,639	2.000	44.575	6,714
T10I4D100K (T)	100,000	11.102	1,276.124	870

Table 8.2 Characteristics of the time databases

D	NT	ALT	AFI	NI	D	NT	ALT	AFI	NI
M_0	812	24.000	295.273	66	M_5	812	24.000	221.454	88
M_1	812	24.000	286.588	68	M_6	812	24.000	216.533	90
M_2	812	24.000	249.846	78	M_7	812	24.000	191.059	102
M_3	812	24.000	282.435	69	M_8	812	24.000	229.271	85
M_4	812	24.000	259.840	75	M_9	816	24.000	227.721	86
E_0	33	7.000	4.620	50	E_5	33	7.000	3.915	59
E_1	33	7.000	5.133	45	E_6	33	7.000	3.500	66
E_2	33	7.000	5.500	42	E_7	33	7.000	3.915	59
E_3	33	7.000	4.812	48	E_8	33	7.000	3.397	68
E_4	33	7.000	3.397	68	E_9	39	7.000	4.550	60
R_0	300	5.590	28.677	68	R_5	300	5.140	26.677	68
R_1	300	5.417	28.000	68	R_6	300	5.510	28.353	68
R_2	300	5.360	27.647	68	R_7	300	5.497	28.338	68
R_3	300	5.543	28.456	68	R_8	300	5.537	28.471	68
R_4	300	5.533	28.382	68	R_9	300	5.477	28.235	68
Rt_0	9,000	11.244	12.070	8,384	Rt_5	9,000	10.856	16.710	5,847
Rt_1	9,000	11.209	12.265	8,225	Rt_6	9,000	11.200	17.416	5,788
Rt_2	9,000	11.337	14.597	6,990	Rt_7	9,000	11.155	17.346	5,788
Rt_3	9,000	11.490	16.663	6,206	Rt_8	9,000	11.997	18.690	5,777
Rt_4	9,000	10.957	16.039	6,148	Rt_9	7,162	11.692	15.348	5,456
B_0	7,482	2.000	5.016	2,983	B_{10}	7,482	2.000	4.573	3,272
B_1	7,482	2.000	4.494	3,330	B_{11}	7,482	2.000	4.895	3,057
B_2	7,482	2.000	5.782	2,588	B_{12}	7,482	2.000	4.636	3,228
B_3	7,482	2.000	4.359	3,433	B_{13}	7,482	2.000	4.805	3,114
B_4	7,482	2.000	4.228	3,539	B_{14}	7,482	2.000	4.192	3,570
B_5	7,482	2.000	4.194	3,568	B_{15}	7,482	2.000	4.656	3,214
B_6	7,482	2.000	3.786	3,952	B_{16}	7,482	2.000	5.379	2,782
B_7	7,482	2.000	3.477	4,304	B_{17}	7,482	2.000	4.863	3,077
B_8	7,482	2.000	4.168	3,590	B_{18}	7,482	2.000	4.654	3,215
B_9	7,482	2.000	4.365	3,428	B_{19}	7,481	2.000	4.953	3,021
T_0	5,000	11.123	64.968	856	T_{10}	5,000	11.113	64.913	856
T_1	5,000	10.987	63.880	860	T_{11}	5,000	11.165	64.988	859
T_2	5,000	11.189	65.128	859	T_{12}	5,000	11.127	64.617	861
T_3	5,000	11.078	64.330	861	T_{13}	5,000	11.089	64.694	857
T_4	5,000	11.003	63.895	861	T_{14}	5,000	11.169	65.088	858
T_5	5,000	11.131	64.867	858	T_{15}	5,000	11.028	64.338	857
T_6	5,000	11.171	64.645	864	T_{16}	5,000	11.132	64.795	859
T_7	5,000	11.075	64.764	855	T_{17}	5,000	11.031	64.661	853
T_8	5,000	11.123	65.121	854	T_{18}	5,000	11.072	64.374	860
T_9	5,000	11.151	64.755	861	T_{19}	5,000	11.090	64.856	855

having high frequency as it is relatively dense. Also, we get many generalized notches having relatively large height as shown in Table 8.3. On the other hand, the items in *retail* is somewhat skewed in the sense that some generalized notches for few items have large height. But, the items in *random-68* and *ecoli* have got more or less

Table 8.3 Top 10 generalized notches in M, E and R databases (according to height)

M (α) at $\beta = 4$				E (α) at $\beta = 2$				R (α) at $\beta = 3$			
Item	Year (sales)	Type	Height	Item	Year (sales)	Type	Height	Item	Year (sales)	Type	Height
116	4(1,489)	u	1,344	42	2(14)	u	13	48	4(48)	u	30
116	7(353)	d	1,136	42	6(1)	d	13	48	5(18)	d	30
114	2(1,131)	u	1,062	35	6(0)	d	12	27	5(19)	d	26
114	4(69)	d	1,062	0	6(10)	d	10	27	8(45)	u	26
56	5(810)	u	796	0	7(20)	u	10	36	6(39)	u	26
56	9(14)	d	796	39	4(11)	u	10	36	8(13)	d	26
67	6(103)	d	704	39	6(1)	d	10	18	8(21)	d	25
94	5(2)	d	650	41	3(11)	u	10	3	5(41)	u	24
94	9(652)	u	650	41	6(1)	d	10	3	7(17)	d	24
1	3(69)	d	644	34	2(2)	d	9	36	2(41)	u	23

Table 8.4 Top 10 generalized notches in Rt, B and T databases (according to height)

Rt (α) at $\beta = 2$				B (α) at $\beta = 3$				T (α) at $\beta = 2$			
Item	Year (sales)	Type	Height	Item	Year (sales)	Type	Height	Item	Year (sales)	Type	Height
41	4(2,617)	u	2,617	333,469	9(582)	u	506	966	4(297)	d	122
41	9(2,355)	u	2,355	333,469	12(76)	d	506	966	6(419)	u	122
0	2(2,331)	d	1,315	333,469	4(151)	d	388	998	2(346)	u	103
48	9(4,544)	u	932	333,469	6(539)	u	388	998	6(243)	d	103
48	4(4,704)	u	711	333,449	9(474)	u	379	966	8(395)	u	93
0	3(2,403)	u	609	333,449	11(95)	d	379	966	10(302)	d	93
0	4(1,794)	d	609	333,449	4(148)	d	372	829	16(431)	u	90
0	7(1,619)	u	571	333,449	6(520)	u	372	966	11(389)	u	87
8,978	2(556)	u	556	110,877	5(353)	u	344	419	4(179)	d	85
0	5(2,089)	u	512	110,877	10(9)	d	344	419	6(264)	u	85

Table 8.5 Top 10 generalized notches in different databases (according to width) at a given α

M (β) at $\alpha = 300$				E (β) at $\alpha = 2$				R (β) at $\alpha = 5$			
Item	Year (sales)	Type	Width	Item	Year (sales)	Type	Width	Item	Year (sales)	Type	Width
56	5(810)	u	8	35	6(0)	d	6	1	5(19)	d	6
11	4(427)	u	7	42	6(1)	d	6	11	8(21)	d	6
13	6(39)	d	7	33	7(1)	d	5	43	5(36)	u	6
16	5(361)	u	7	41	3(11)	u	5	6	3(32)	u	5
67	6(103)	d	7	42	2(14)	u	5	11	4(32)	u	5
94	5(2)	d	7	49	6(1)	d	5	18	8(21)	d	5
98	2(404)	u	7	52	4(1)	d	5	47	5(31)	u	5
11	8(32)	d	6	40	4(8)	u	4	50	4(18)	d	5
52	6(703)	u	6	45	2(8)	u	4	50	8(35)	u	5
53	6(109)	d	6	45	5(1)	d	4	53	6(40)	u	5

Table 8.6 Top 10 generalized notches in different databases (according to width) at a given α

Rt (β) at $\alpha = 20$				B (β) at $\alpha = 50$				T (β) at $\alpha = 5$			
Item	Year (sales)	Type	Width	Item	Year (sales)	Type	Width	Item	Year (sales)	Type	Width
9,823	7(30)	u	9	112,551	9(6)	d	10	673	8(71)	d	8
2,046	4(133)	u	8	335,213	10(4)	d	10	651	11(82)	u	8
3,321	7(24)	u	8	112,339	5(224)	u	9	524	4(29)	u	8
411	4(32)	u	8	335,185	4(130)	u	9	283	15(229)	u	7
3,121	4(0)	d	8	112,407	5(107)	u	9	487	15(139)	d	7
2,919	6(32)	u	8	335,181	5(54)	u	9	336	13(42)	d	7
103	7(267)	u	7	335,213	5(54)	u	9	523	11(117)	u	7
855	3(118)	u	7	335,177	5(51)	u	9	658	14(88)	d	7
1,659	3(116)	u	7	110,877	5(353)	u	8	733	16(63)	u	7
976	6(55)	d	7	110,315	11(310)	u	8	807	10(29)	u	7

uniform distribution. Many generalized notches in these two databases have similar height. Unlike *mushroom* and *retail*, *ecoli* and *random-68* are smaller in size and contain items with lesser variations. In these two databases the maximum heights are 13 and 30, respectively. With respect to width of a generalized notch, we have obtained similar characteristics. In *mushroom* and *retail*, the generalized notches are wider than that of other two databases. These facts are quite natural, since *mushroom* and *retail* are bigger than *random-68* and *ecoli*. The variation of sales over the years for an item in *mushroom* and *retail* is higher. Also, we observe that many upward generalized notch is followed by a downward generalized notch and vice versa. This is because of the fact that two consecutive different types (a 'u' type followed by a 'd' type or a 'd' type followed by an 'u' type) generalized notches share a common leg. For example, a 'u' type generalized notch at year 4 is followed by a 'd' type generalized notch at year 7, for item 116 in *mushroom*. Similarly, a 'd' type generalized notch at year 6 is followed by a 'u' type generalized notch at year 7, for item 0 in *ecoli*. Also, we observe that some items have both long height and long width. For example, item 56 has height 796 and width 8. These values are significantly high as compare to other items in the time databases. Also, this is true for item 67. In *BMS-WebView-1* database items 333469 and 110877 have maximum variation. Therefore, only these two items are appearing among top ten generalized notches and their heights vary from 344 to 506. Similarly, items 966, 998 and 419 in *T10I4D100K* share common legs and they have more variations. From Table 8.6 one could conclude that generalized notches are sharper in *BMS-WebView-1* as compared to *T10I4D100K* (Table 8.7).

We have also reported an execution time with respect to the number of data sources. We observe in Figs. 8.2, 8.3, 8.4, 8.5, 8.6 and 8.7 that the execution time increases as the number of databases increases. The size of each input database generated from *mushroom*, *retail*, *BMS-WebView-1* and *T10I4D100K* are significantly larger than that of *ecoli*. As a result, we observe a steeper graph in Figs. 8.3 and 8.6. We have fixed α for *ecoli* and *random-68* at lower level, since variation of

Table 8.7 Significant years in different databases at a given δ

M $\delta = 0.04$ Significant years	E $\delta = 0.05$ Significant years	R $\delta = 0.04$ Significant years	Rt $\delta = 0.0004$ Significant years	B $\delta = 0.0006$ Significant years	T $\delta = 0.003$ Significant years
6	6	8	7	4	6
4	4	7	6	2	
3		5	4		

Fig. 8.2 Execution time versus the number of databases (*mushroom* at $\alpha = 50, \beta = 3$)

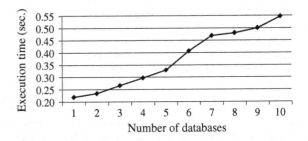

Fig. 8.3 Execution time versus the number of databases (*ecoli* at $\alpha = 3$, $\beta = 2$)

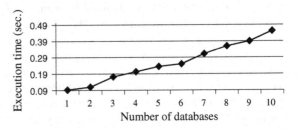

Fig. 8.4 Execution time versus the number of databases (*random-68* at $\alpha = 3, \beta = 2$)

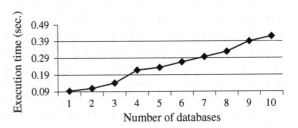

Fig. 8.5 Execution time versus the number of databases (*retail* at $\alpha = 20$, $\beta = 2$)

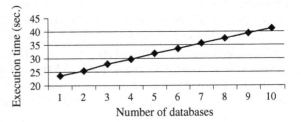

Fig. 8.6 Execution time versus the number of databases (*BMS-WebView-1* at $\alpha = 30$, $\beta = 4$)

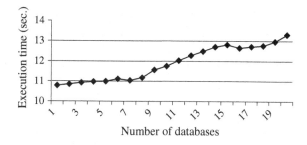

Fig. 8.7 Execution time versus the number of databases (*T10I4D100K* at $\alpha = 30$, $\beta = 4$)

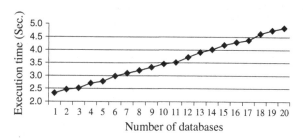

Fig. 8.8 Number of interesting icebergs versus height (α) for *mushroom* ($\beta = 3$)

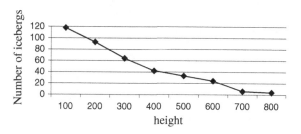

Fig. 8.9 Number of interesting icebergs versus width (β) for *mushroom* ($\alpha = 50$)

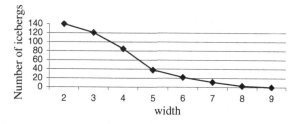

frequencies of an item is lesser. We observe that the execution time of *retail* is significantly larger than for other databases, since each of the time databases is comparatively larger. In Figs. 8.7 and 8.8 we have considered same α and β for *BMS-WebView-1* and *T10I4D100K*, respectively. But execution time of *BMS-WebView-1* is significantly larger than the one reported for *T10I4D100K*.

In Figs. 8.8, 8.9, 8.10, 8.11, 8.12, 8.13, 8.14, 8.15, 8.16, 8.17, 8.18, and 8.19 we have presented how the number of interesting icebergs decreases with respect to the increase of the values of α and β. In *mushroom*, *ALT* and *AFI* are higher as

Fig. 8.10 Number of interesting icebergs versus height (α) for *ecoli* ($\beta = 2$)

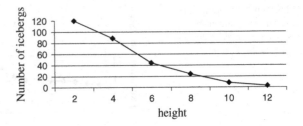

Fig. 8.11 Number of interesting icebergs versus width (β) for *ecoli* ($\alpha = 3$)

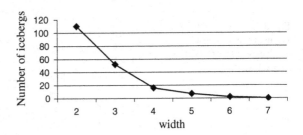

Fig. 8.12 Number of interesting icebergs versus height (α) for *random-68* ($\beta = 2$)

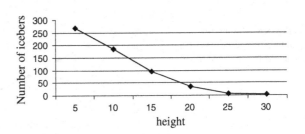

Fig. 8.13 Number of interesting icebergs versus width (β) for *random-68* ($\alpha = 3$)

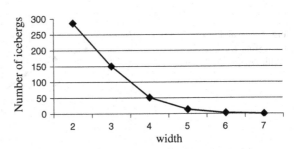

Fig. 8.14 Number of interesting icebergs versus height (α) for *retail* ($\beta = 2$)

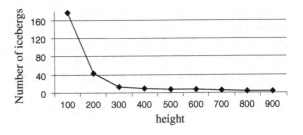

Fig. 8.15 Number of interesting icebergs versus width (β) for *retail* (α = 20)

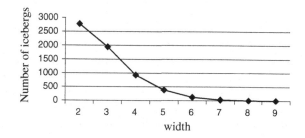

Fig. 8.16 Number of interesting icebergs versus height (α) for *BMS-WebView-1* (β = 4)

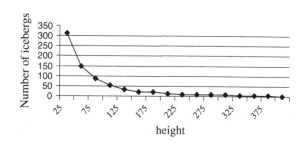

Fig. 8.17 Number of interesting icebergs versus width (β) for *BMS-WebView-1* (α = 30)

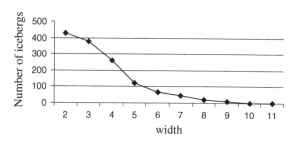

Fig. 8.18 Number of interesting icebergs versus height (α) for *T10I4D100K* (β = 5)

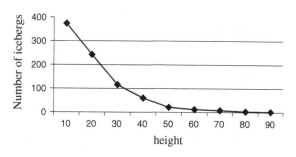

Fig. 8.19 Number of interesting icebergs versus width (β) for *T10I4D100K* (α = 30)

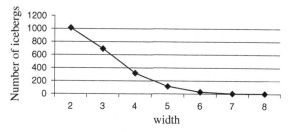

compared to other databases. We start changing the values α proceeding from 100 (Fig. 8.8). Initially, the number of icebergs decreases significantly. Afterwards, the decrease is not so significant. As shown In Fig. 8.9, the number of icebergs decreases significantly when the width remains lower. Afterwards, the number of icebergs decreases slowly.

ALT is smaller for *ecoli* and *random-68*. As a result, the height of an iceberg remains smaller. We start α from 2 and 5 for *ecoli* and *random-68*, respectively (Figs. 8.10 and 8.12). We have not discovered any interesting icebergs for the width greater than or equal to 7.

8.10 Conclusions

The study of temporal patterns in time-stamped databases is an important research and application-oriented issue. Many interesting patterns have been discovered in transactional as well as in time series databases. In this chapter, we have proposed definitions of different patterns in time-stamped databases. First, we have introduced the notion of notch in a sales series of an item. Based on this pattern, we have introduced two more patterns viz., generalized notch and iceberg notch, present in sales series of an item. Iceberg notch represents a special sales pattern of an item over time. It could be considered as an exceptional pattern in time-stamped databases. The investigations of such patterns could be important to understand the purchasing behaviour of customers. They also help identifying the reasons for such a behaviour. We have designed an algorithm to extract icebergs in time-stamped databases and presented experimental results for real-world and synthetic time-stamped databases.

References

Adhikari A, Rao PR (2008) Mining conditional patterns in a database. Pattern Recogn Lett 29(10):1515–1523

Adhikari J, Rao PR (2010) Measuring influence of an item in a database over time. Pattern Recogn Lett 31(3):179–187

Adhikari J, Rao PR, Adhikari A (2009) Clustering items in different data sources induced by stability. Int Arab J Inf Technol 6(4):394–402

Adhikari J, Rao PR, Pedrycz W (2011) Mining icebergs in time-stamped databases. In: Proceedings of Indian International Conferences on Artificial Intelligence, pp 639–658

Agrawal R, Srikant R (1994) Fast algorithms for mining association rules. In: Proceedings of the international conference on very large databases, pp 487–499

Agrawal R, Srikant R (1995) Mining sequential patterns. In: Proceedings of international conference on data engineering, pp 3–14

Agrawal R, Psaila G, Wimmers EL, Zaït M (1995) Querying shapes of histories. In: Proceedings of 21st VLDB Conference, pp 502–514

Agrawal R, Imielinski T, Swami A (1993) Mining association rules between sets of items in large databases. In: Proceedings of ACM SIGMOD conference on management of data, pp 207–216

Böttcher M, Hoppner F, Spiliopoulou M (2008) On exploiting the power of time in data mining. SIGKDD Explor 10(2):3–11

Box G, Jenkins G, Reinsel G (2003) Time series analysis, 3rd edn. Pearson Education, NJ

Brockwell J, Richard D (2002) Introduction to time series and forecasting. Springer, Berlin

Castellana L, D'Addabbo A, Pasquariello G (2007) A composed supervised/unsupervised approach to improve change detection from remote sensing. Pattern Recogn Lett 28(4):405–413

Fink E, Gandhi HS (2007) Important extrema of time series. In: Proceedings of the IEEE international conference on systems, man and cybernetics, pp 366–372

Frequent Itemset Mining Dataset Repository, http://fimi.cs.helsinki.fi/data

Han J, Kamber M (2001) Data mining: concepts and techniques. Morgan Kauffmann, Burlington

Han J, Pei J, Yiwen Y (2000) Mining frequent patterns without candidate generation. In: Proceedings of ACM SIGMOD international conference management of data, pp 1–12

Hong TP, Wu YY, Wang SL (2009) An effective mining approach for up-to-date patterns. Expert Systems with Applications (36):9747–9752

Keogh E (1997) A fast and robust method for pattern matching in time series databases. In: Proceedings of 9th international conference on tools with AI (ICTAI), pp 578–584

Keogh E, Lonardi S, Chiu B (2002) Finding surprising patterns in a time series database in linear time and space, In: Proceedings of KDD international conference, pp 23–26

Keogh E, Lin J, Lee SH, Herle HV (2006) Finding the most unusual time series subsequence: algorithms and applications. Knowl Inf Syst 11(1):1–27

Leonard M., Wolfe B (2005) Mining transactional and time series data. In: Proceedings of the SUGI 30, pp 18–24

Li D, Deogun JS (2005) Discovering partial periodic sequential association rules with time lag in multiple sequences for prediction. Foundations of Intelligent Systems LNCS 3488/2005:1–24

Li Y, Ning P, Wang XS, Jajodia S (2003) Discovering calendar-based temporal association rules. Data and Knowledge Engineering 44(2):193–218

Lomet DB, Hong M, Nehme RV, Zhang R (2008) Transaction time indexing with version compression. Proc VLDB Endowment 1(1):870–881

Mahanta AK, Mazarbhuiya FA, Baruah HK (2008) Finding calendar-based periodic patterns. Pattern Recogn Lett 29(9):1274–1284

Perng CS, Wang H, Zhang SR, Parker DS (2000) Landmarks: A new model for similarity-based pattern querying in time-series databases. In: Proceedings of 16th International Conference on Data Engineering, pp 33–42

Pratt K, Fink E (2002) Search for patterns in compressed time series. Int J Image Graph 2(1):89–106

Roddick JF, Spillopoulou M (1999) A Bibliography of temporal, spatial and spatio-temporal data mining research. ACM SIGKDD Explor 1(1):34–38

Singh S, Stuart E (1998) A pattern matching tool for time series forecasting. In: Proceedings of 14th international conference on pattern recognition, pp 103–105

Tsay RS (1989) Identifying multivariate time series models. J Time Ser Anal 10(4):357–372

UCI ML Repository, http://www.ics.uci.edu/~mlearn/MLSummary.html

Wang L, Chang ES, Li H (2010) A tree-construction search approach for multivariate time series motifs discovery. Pattern Recogn Lett 31(9):869–875

Wu X, Zhang C, Zhang S (2004) Efficient mining of both positive and negative association rules. ACM Trans Inf Syst 22(3):381–405

Chapter 9
Mining Calendar-Based Periodic Patterns in Time-Stamped Data

A large class of problems is concerned with temporal data. Identifying temporal patterns in these datasets is a fully justifiable as well as an important task. Recently, researchers have reported an algorithm for finding calendar-based periodic pattern in a time-stamped data and introduced the concept of certainty factor in association with an overlapped interval. In this chapter, we have extended the concept of certainty factor by incorporating support information for effective analysis of overlapping intervals. We have proposed a number of improvements of the algorithm for identifying calendar-based periodic patterns. In this direction we have proposed a hash based data structure for storing and managing patterns. Based on this modified algorithm, we identify full as well as partial periodic calendar-based patterns. We provide a detailed data analysis incorporating various parameters of the algorithm and make a comparative analysis with the existing algorithm, and show the effectiveness of our algorithm. Experimental results are provided on both real and synthetic databases.

9.1 Introduction

A large amount of data being collected every day exhibits a temporal connotation. For example, databases which originate from transactions in a supermarket, logs in a network, transactions in a bank, and events related to manufacturing industry are all inherently related to time. Data mining techniques could also be applied to these databases to discover various temporal patterns to understand the behavior of customers, markets, or monitored processes in different points of time. Temporal data mining is concerned with the analyses of data to find out patterns and regularities from a set of temporal data. In this context, sequential association rule (Agrawal and Srikant 1995), periodical association rule (Li and Deogun 2005), calendar association rule (Li et al. 2003), calendar-based periodic pattern (Mahanta et al. 2008), and up-to-date pattern (Hong et al. 2009) are some interesting temporal patterns reported in the recent time.

A. Adhikari et al., *Data Analysis and Pattern Recognition in Multiple Databases*,
Intelligent Systems Reference Library 61, DOI: 10.1007/978-3-319-03410-2_9,
© Springer International Publishing Switzerland 2014

For effective management of business activities, we often wish to discover knowledge from time-stamped data. There are several important aspects of mining time-stamped data including trend analysis, similarity search, forecasting and mining of sequential and periodic patterns. In a database from a retail store, the sales of ice cream in summer and the sales of blanket in winter should be higher than those of the other seasons. Such seasonal behaviour of specific items can only be discovered when a proper window size is chosen for the data mining process (Roddick and Spiliopoulou 2002). A supermarket manager may discover that turkey and pumpkin pie are frequently sold together in November in every year. Discovering such patterns may reveal interesting information that can be used for understanding the behaviour of customers, markets or monitored processes in different time periods. However, these types of seasonal patterns cannot be discovered by traditional non-temporal data mining approaches that treat all the data as one large segment with no attention paid to utilizing the time information of the transactions. If one looks into the entire dataset rather than the transactions that occur in November, it is likely that one will not be able to discover the pattern of turkey and pumpkin pie since the overall support for them will be evidently low. In general, a time-stamped database might exhibit some periodic behaviours. Length of a period might vary from one context to another context. For example, in case of sales of ice cream, the basic time interval could be of 3 months, since in many regions March, April and May together is considered as summer. Also, in case of sales of blanket, the basic time interval could be considered from November to February in every year. In addition, in many business applications, one might be interested in quarterly patterns over the years, where length of the period is equal to 3 months. A large amount of data is collected every day in the form of event time sequences. These sequences are valuable sources to analyze not only the frequencies of certain events, but also the patterns with which these events occur. For example, from data consisting of web clicks one may discover that a large number of web browsers who visit www.washingtonpost.com in morning hours also visit www.cnn.com. Using such information one can group users as daily morning users, daily evening users, weekly users, etc. This information might be useful for communicating to the users. Temporal patterns in a stock market, such as whether certain months, days of the week, time periods or holidays provide better returns than other time periods have received particularly a large amount of attention. Due to the presence of various types of applications in many fields, periodic pattern mining is an interesting area of study.

Mahanta et al. (2008) used set superimposition (Baruah 1999) to find the membership value of each fuzzy interval. The concept of set *superimposition* is defined as follows. If set A is superimposed over set B or set B is superimposed over set A then set superimposition operation can be expressed as $A\,(S)\,B = (A - B)$ $(+)\,(A \cap B)^{(2)}\,(+)\,(B - A)$, where (S) denotes the set superimposition operation. Here, the elements of $(A \cap B)^{(2)}$ are the elements of $(A \cap B)$ represented twice and $(+)$ represents union of disjoint sets. Authors have also designed an algorithm for mining calendar-based periodic patterns. While applying this concept authors have assumed intervals with equal membership grade, and accordingly the concept of certainty factor has been proposed for each sub-interval. Certainty factor of an interval over different time periods expresses the likelihood of reporting the pattern in that particular interval. If two intervals overlap then the certainty factor is more for the overlapped region than the non-overlapped region. When two intervals are superimposed, authors have assumed 1/2 membership grade for each interval. After superimposition, the membership grade fuzzy membership value for the overlapped region becomes 1. The membership grade for non-overlapped region remains 1/2. But these two intervals may have different supports for the pattern. The *support* (Agrawal et al. 1993) of a pattern represents a fraction of transactions containing the pattern. A pattern is *frequent* if its support is greater than equal to a user-defined threshold, *minsupp*. The certainty factor and support of a pattern in an interval are two different concepts. For an effective analysis of overlapped regions, these two concepts need to be introduced along with an overlapped region. Thus, in this chapter we propose an extended analysis of superimposed intervals. The main weak point of the aforementioned paper is that the concept of set superimposition is not necessary in the proposed algorithm. Therefore, we have proposed a modified algorithm for identifying full as well as partial calendar-based periodic patterns. We have also improved the proposed algorithm by introducing a hash based data structure for storing relevant information associated with intervals. In addition, we have suggested some other improvements in the proposed algorithm. Before concluding this section, we give an example of a time-stamped database that will be used for providing illustrative examples on various concepts.

Example 9.1 Consider the following database *D* of transactions. Each record contains items purchased as well as the date of the transaction (Table 9.1).

We have omitted the time of a transaction, since our data analysis is not associated with the time component of a transaction. We will refer to this database from time to time for the purpose of illustrating various concepts.

The chapter is organized as follows. We discuss related work in Sect. 9.2. In Sect. 9.3, we have discussed calendar-based periodic patterns and proposed an extended certainty factor of an interval. We have designed an algorithm for identifying calendar-based periodic patterns in Sect. 9.4. Experimental results are provided in Sect. 9.5. We conclude the chapter in Sect. 9.6. •

Table 9.1 A sample time-stamped database

Time-stamp	Items	Time-stamp	Items	Time-stamp	Items
29/03/1990	a, b, c	07/04/1992	a, c, e, g, h	17/04/1993	a, c, f
06/04/1990	a, c, e	12/04/1992	c, e	06/04/1994	a, b, c, d
21/04/1990	a, d	14/04/1992	c, e, f	10/04/1994	g, h
25/04/1990	a, c, d	19/04/1992	f, g	13/04/1994	a, g
06/03/1991	a, c	04/03/1993	a, c	18/04/1994	g, h, i
12/03/1991	a, c, e	09/03/1993	a, c, g	20/04/1994	a, c, e, f
19/04/1991	f, g	01/04/1993	c, h, i		
03/03/1992	a, c, d	07/04/1993	c, d		

9.2 Related Work

A calendar time expression is composed of calendar units in a specific calendar and represents different time features, such as an absolute time interval and a periodic time over a specific time period. A calendar-based periodic pattern is associated with time hierarchy for calendar years. In this study, we have dealt with calendar dates over the years.

Verma et al. (2005) have proposed an algorithm H-Mine, where a header table H is created separately for each interval. Each frequent item entry has three fields viz., item-id, support count and a hyper-link. In order to deal with the patterns in time-stamped databases we have proposed a hash-based data structure where at the first index level we store distinct years that appear in the transactions. Then we keep an array of pointers corresponding to every year in the index table. The k-th pointer of this array points to tables containing interesting itemsets of size k.

Lee et al. (2009) have proposed two data mining systems for discovering fuzzy temporal association rules and fuzzy periodic association rules. The mined patterns are expressed in fuzzy temporal and periodic association rules that satisfy the temporal requirements specified by the user. In the proposed algorithm the mined patterns are dependent on user inputs such as maximum gap between two intervals and minimum length of an interval.

Li et al. (2003) proposed two classes of temporal association rules, temporal association rules with respect to precise match and temporal association rules with respect to fuzzy match, to represent regular association rules along with their temporal patterns. Our work differs from it, since we identify frequent itemsets along with the associated intervals. Then we use match ratio to determine whether a pattern is full periodic or partial periodic. Subsequently, Zimbrao et al. (2002) reported a similar work. Authors incorporate multiple granularities of time intervals from which both cyclic and user-defined calendar patterns can be achieved. Ale and Rossi (2000) proposed an algorithm to discover temporal association rules. In this algorithm, support of an item is calculated only during its lifespan. In the proposed work we compute and store supports of itemsets when they satisfy the requirements of the user.

Lee et al. (2006) have proposed a technique for mining partial multiple periodic patterns without redundant rules. Without mining every period, authors checked the necessary period and used this information to do further mining. Instead of considering the whole database, the information needed for mining partial periodic patterns is transformed into a bit vector that can be stored in a main memory. This approach needs to scan the database at the most two times. Our approach extracts both partial and full periodic patterns together by scanning the database repeatedly to find the higher-level patterns as done using apriori algorithm (Agrawal and Srikant 1994).

In the context of support definition, Kempe et al. (2008) have proposed a new support definition that counts the number of pattern instances, handles multiple instances of a pattern within one interval sequence and allows time constraints on a pattern instance.

Lee et al. (2002) have proposed a new temporal data mining technique that can extract temporal interval relation rules from temporal interval data by using Allen's theory (Allen 1983). Authors designed a preprocessing algorithm for generalization of temporal interval data. Also, authors have proposed an algorithm for discovering a temporal interval relation. Although there are thirteen different types of relations between two intervals, in our work we have focused on only overlapped intervals to find locally frequent itemsets of larger size and detect periodicity of patterns.

Ozden et al. (1998) proposed a method of finding patterns having periodic nature where the period has to be specified by the user. Han et al. (1999) proposed several algorithms for mining partial periodic patterns by exploring some interesting properties such as the apriori property and the max-subpattern hit set property by shared mining of multiple periods.

9.3 Calendar-Based Periodic Patterns

In Sects. 9.1 and 9.2, we have presented some important applications of calendar-based periodic patterns. A calendar-based periodic pattern is dependent on the schema of a calendar. There are various ways one could define the schema of a calendar. We assume that the schema of calendar-based pattern is based on day, month and year. This schema is also useful to determine weekly-based pattern, since first seven days of any month correspond to the first week, days 8–14 of any month correspond to the second week, and so on. Thus, one can have several types of calendar-based periodic patterns viz., daily, weekly, monthly and yearly. Based on a schema, some examples of calendar patterns are given as follows: every day of January, 1999; every 16-th day of January in each year; second week of every month. Again, each of these periodic patterns could be of two types viz., partially periodic pattern and full periodic pattern. A problem related to periodicity could be of finding patterns occurring at regular time intervals. Thus it emphasizes on two aspects viz., pattern and interval.

A calendar pattern refers to a market cycle that repeats periodically on a consistent basis. Seasonality could be a major force in a marketplace. While calendar patterns are based on a framework of multiple time granularities viz., day, month and year, but the periodic patterns are defined in terms of a single granularity. Here patterns are dependent on the lifespan of an item in a database. Lifespan of an item (x) is a pair $(x, [t_1, t_2])$, where t_1 and t_2 denote the time that the item x appears in the database for the first time and last time, respectively. The problem of periodic pattern mining can be categorized into two types. One is full periodic pattern mining, where every point in time granularity (Bettini et al. 2000) contributes to a cyclic behavior of the pattern. The other and more general one is called partial periodic pattern mining, which specifies the behavior of the pattern at some but not all points of time granularity in the database. Partial periodicity is a looser form of periodicity than full periodicity, and it also occurs more commonly in a real world database. A pattern is associated with a real number m $(0 < m < 1)$, called match ratio (Li et al. 2003) that reveals that a pattern holds with respect to fuzzy match satisfying at least 100 m % of the time intervals. Match ratio is an important measure which determines whether a calendar-based pattern could be full periodic or partial periodic. When the match ratio is equal to 1 then it is a full periodic pattern. In case of partial periodic pattern the match ratio lies between 0 and 1. While finding yearly periodic patterns, Mahanta et al. (2008) have proposed match ratio in somewhat a different way. Authors have proposed match ratio as the number of intervals is divided by the number of years in the lifespan of the pattern for the purpose of mining yearly pattern. It might be difficult to work with this definition, since a mining algorithm returns itemsets and their intervals. A mining algorithm might not be concerned with reporting the first and last appearances of an itemset. Therefore, we will follow the definition proposed by Li et al. (2003).

We have discussed the concept of certainty factor in Sect. 9.1. Also we have noticed that the analysis of overlapped region using certainty factor might not be sufficient. Therefore, we propose its extension.

9.3.1 Extending Certainty Factor

The concept of certainty factor is based on the concept of set superimposition. If we are interested in yearly patterns, during the analysis of superimposed intervals the year component is ignored. We explain here the concept of set superimposition using the following example.

Example 9.2 Consider the database of Example 9.1. Itemset $\{a, c\}$ is present in 3 out of 4 transactions in the intervals [29/03/1990–25/04/1990]. Also, $\{a, c\}$ is present in 2 out of 3 transactions in the intervals and [06/03/1991–19/04/1991]. Therefore, $\{a, c\}$ is frequent in these intervals at minimum support level of 0.66. These two intervals are being superimposed where each of these intervals has fuzzy set membership value 1/2. The overlapped area of these two intervals is [29/03–19/04]. Based on the concept of set superimposition, an itemset reported in a non-overlapped region has the fuzzy set membership value 1/2. But, an itemset reported in the overlapped interval [29/03–19/04] has the set membership value equal to $1/2 + 1/2 = 1$. •

For the purpose of mining periodic patterns, Mahanta et al. (2008) have proposed the use of the certainty factor. It is based on a set of overlapped intervals corresponding to a pattern occurring on a periodic basis. For example, one might be interested in identifying yearly periodic patterns in a database. Authors have considered all the intervals having equal membership grade. For example, if n intervals are superimposed then every interval has $1/n$ equal fuzzy membership grade and in an overlapped area the membership value will be added. The certainty of the pattern in the overlapped interval is more than the certainty in the other intervals. Let $[t_1, t_1']$ and $[t_2, t_2']$ be two overlapped intervals where a pattern X gets reported with certainty value 1/2. When the two intervals are superimposed the certainty factors of X associated with the various subintervals are given as follows:

$$\left[t_1, t_1'\right]^{1/2}(S)\left[t_2, t_2'\right]^{1/2} = \left[t_1, t_2\right)^{1/2}\left[t_2, t_1'\right]^{1}\left(t_1', t_2'\right]^{1/2} \tag{9.1}$$

The notion of certainty factor seems to be an important contribution made by the authors. It represents the certainty of reporting a pattern in an interval by considering a sequence of periods. For example, we might be interested in knowing the certainty of pattern $\{a, c\}$ in the month of April with respect to the database in Example 9.1. It is an important statistical evidence of a pattern being in an interval over a sequence of years (periods). For example, one could say that the evidence of the pattern $\{a, c\}$ is certain in the month of April when the years viz., 1990, 1991, 1992 and 1993 are considered. But the concept of certainty factor does not convey the information regarding the frequency of a pattern in an overlapped region. In addition, it gives equal importance to all the intervals by considering them as equal fuzzy grade intervals. From the perspective of the evidence of a pattern, such assumption might be realistic. But from the perspective of the depth of evidence, such concept might not be sufficient. Thus, we propose an extension to the concept of certainty factor. In the proposed extension, we incorporate the information regarding support of a pattern in an interval. There are many ways one could keep the information regarding support. In Example 9.1, there are four overlapping intervals corresponding to the pattern $\{a, c\}$. There exists a region where all the intervals are overlapped, while some regions may not be overlapped at all. Apart from the certainty factor of a region, one could also keep the support information of the pattern in that interval. In general, a region could be overlapped by all intervals. Let there be n supports of a pattern corresponding to n intervals. Then the question comes to our mind, how to keep the support information of the pattern for n intervals. The answer to this question might not be agreeable to all. One might be interested in keeping the average support of the pattern along with the certainty factor for that interval. Some of us might be interested in keeping information regarding the minimum and maximum of n supports. In an extreme case, one might be interested in keeping all the n supports of the pattern corresponding to n intervals. Let us consider that we are interested in yearly pattern. Let the lifespan of a pattern be 40 years. Then one has to keep a maximum of forty supports corresponding to an overlapped region. It might not be realistic to maintain all the forty supports. Let $s\text{-}info(X, [t_1, t_2])$ be the support information of the pattern X for the interval $[t_1, t_2]$. Let a pattern X be frequent in time intervals $[t_i, t_i']$, $i = 1, 2, \ldots, n$. Each of these intervals is taken from a different period of time such that $\cap_{i=1}^{n}[t_i, t_i'] \neq \varphi$. In Example 9.1, patterns $\{a\}$, $\{c\}$ and $\{a, c\}$ get reported in the month of April in every year. By generalizing (9.1), the certainty factor of X in overlapped regions could be obtained as follows:

$$\left[t_1, t_1'\right]^{1/n}(S)\left[t_2, t_2'\right]^{1/n}(S)\ldots(S)\left[t_n, t_n'\right]^{1/n} = \left[t^{(1)}, t^{(2)}\right)^{1/n}\left[t^{(2)}, t^{(3)}\right)^{2/n}\left[t^{(3)}, t^{(4)}\right)^{3/n}\ldots\left[t^{(r)}, t^{(r+1)}\right)^{r/n}\ldots$$
$$\times \left[t^{(n)}, t'^{(1)}\right]^{1}\left(t'^{(1)}, t'^{(2)}\right]^{n-1/n}\ldots\left(t'^{(n-2)}, t'^{(n-1)}\right]^{2/n}\left(t'^{(n-1)}, t'^{(n)}\right]^{1/n}$$

$$\tag{9.2}$$

where $\{t^{(i)}\}_{i=1}^{n}$ is the sequence obtained from $\{t_i\}_{i=1}^{n}$ by sorting in ascending order and $\{t'^{(i)}\}_{i=1}^{n}$ is obtained from $\{t'_i\}_{i=1}^{n}$ by sorting in ascending order. We propose an extended certainty factor of X in the above overlapped intervals as follows:

When X is reported in $[t^{(n)}, t'^{(1)}]$ then the certainty value is 1 with support information $s\text{-}info(X, [t^{(n)}, t'^{(1)}])$. But, the certainty value of X for the outside of $[t^{(1)}, t'^{(n)}]$ is 0 with support information 0. When X is reported in $[t^{(r-1)}, t^{(r)})$, then the certainty value is $(r-1)/n$ with support information $s\text{-}info(X, [t^{(r-1)}, t^{(r)}))$, for $r = 2,$ 3, ..., n. Otherwise, the certainty value of X for $(t'^{(r-1)}, t'^{(r)}]$ is $(n-r+1)/n$ with support information $s\text{-}info(X, (t'^{(r-1)}, t'^{(r)}])$, for $r = 3, 4, ..., n$.

Suppose we are interested in identifying yearly periodic patterns. So each time interval is taken from a year. From the perspective of n years, the pattern X gets reported in every year in the interval $[t^{(n)}, t'^{(1)}]$. So the certainty of X is 1 (the highest) in this interval. But, X is not frequent pattern outside of $[t^{(1)}, t'^{(n)}]$. Therefore, from the perspective of all the years the certainty of X is 0 (lowest) outside of the interval. The certainty factor also provides the information regarding how many intervals are overlapped on a sub-interval. For example, if the certainty factor of a sub-interval is 2/5, for given five intervals, then two intervals are overlapped on the sub-interval. On the other hand, $s\text{-}info$ provides the information regarding degree of frequency of X in an interval. To illustrate the above concept we consider the following example.

Example 9.3 The purpose of this example is to explain the proposed concept of extended certainty factor stated above. Let the years 1980, 1981, 1982 and 1983 be of our interest. We would like to check whether the pattern X is yearly periodic. Assume that the mining algorithm has reported X as frequent in the time intervals $[t_1, t'_1]$, $[t_2, t'_2]$, $[t_3, t'_3]$ and $[t_4, t'_4]$ for the years 1980, 1981, 1982 and 1983, respectively. Also, let the supports of X in $[t_1, t'_1]$, $[t_2, t'_2]$, $[t_3, t'_3]$ and $[t_4, t'_4]$ be 0.2, 0.15, 0.16 and 0.12, respectively. Based on the proposed extended concept, we wish to analyze the time interval $[t_4, t'_4]$ by overlapping these intervals corresponding to the 4 years. The overlapped intervals are depicted in Fig. 9.1.

While computing support information we use here the range measure for a set of values. One could use another support information depending on the requirement. An analysis of the overlapped intervals corresponding to X is presented in Table 9.2. Certainty of a sub-interval is based on the number of intervals overlapped with it. For example, $[t_1, t_2)$ comes with certainty of 1/4, since there is only one interval out of four intervals.

Fig. 9.1 Overlapped intervals for finding yearly pattern X

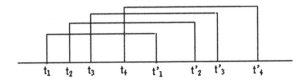

Table 9.2 An analysis of the overlapped intervals for finding yearly pattern X

Interval	Certainty factor	s-info	Interval	Certainty factor	s-info
$[t_1, t_2)$	1/4	0.2–0.2	$(t'_1, t'_2]$	3/4	0.12–0.15
$[t_2, t_3)$	1/2	0.15–0.2	$(t'_2, t'_3]$	1/2	0.12–0.16
$[t_3, t_4)$	3/4	0.15–0.2	$(t'_3, t'_4]$	1/4	0.12–0.12
$[t_4, t'_1]$	1	0.12–0.2			

Here *s-info* corresponding to interval $[t_3, t_4)$ represents the fact that the maximum and minimum supports of overlapped intervals are 0.2 and 0.15 respectively. •

Certainty factor and support information are not the same. They represent two different aspects of a pattern in an interval. Certainty factor is normally associated with multiple time intervals. It expresses the likelihood of reporting a pattern in a sub-interval of the multiple overlapped intervals. But the concept of support is associated with a single time-interval. It is defined as the fraction of the transactions containing the pattern in a time-interval. Thus, for an effective analysis of a superimposed interval both the certainty factor and support information are needed in association with an interval.

9.3.2 Extending Certainty Factor With Respect to Other Intervals

In Fig. 9.1, we have shown four intervals overlapped corresponding to four different years. But in reality the scenario could be different. For four intervals, there may exist different combinations of overlapped intervals. But, whatever may be the case, the certainty factor of a sub-interval depends on the number of intervals overlapped in that sub-interval and *s-info* depends on the supports of the pattern in the intervals that are being overlapped on a sub-interval. Let us consider a sub-interval $[t, t']$, where m out of n intervals are overlapped on $[t, t']$. Based on certainty factor (Mahanta et al. 2008), we propose an extended certainty factor as follows:

When X is reported in $[t, t']$, then the certainty value is m/n with support information $s\text{-}info(X, [t, t'])$, where $s\text{-}info(X, [t, t'])$ is based on supports of X in the m intervals overlapped on $[t, t']$. We illustrate this issue with the help of Example 9.4. Before that, we present a few definitions related to overlapped intervals. Let *maxgap* be the user-defined maximum gap (time units) between current time-stamp of a pattern and the time-stamp of the pattern when it was last seen. If the gap between current time-stamp of a pattern and the time-stamp of the pattern when it was last seen is greater than *maxgap* then a new interval is formed for the pattern with the current time-stamp as the start of the interval. Also, the previous interval of the pattern was ended when it was seen last time. Let *mininterval* be the minimum period length of a time interval. Each interval should be of sufficient

Fig. 9.2 Overlapped
intervals for finding yearly
pattern {a, c}

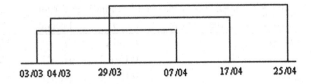

03/03 04/03　　　29/03　　　07/04　　　17/04　　　25/04

Table 9.3 An analysis of the time interval [03/03–25/04] for finding yearly pattern {a, c}

Interval	Certainty	s-info	Interval	Certainty	s-info
[03/03–04/03)	1/5	1.0–1.0	(07/04–17/04]	2/5	0.6–0.75
[04/03–29/03)	2/5	0.6–0.75	(17/04–25/04]	1/5	0.75–0.75
[29/03–07/04]	3/5	0.6–1.0			

length, otherwise a pattern appearing once in a transaction also becomes frequent in an interval. If two intervals are overlapped and the length of the overlapped region exceeds *mininterval* then the overlapped region could be interesting.

Example 9.4 We refer to the database of Example 9.1. Let the value of *maxgap* be 40 days. Then pattern {a, c} gets reported in the following intervals: [29/03/ 1990–25/04/1990], [06/03/1991–12/03/1991], [03/03/1992–07/04/1992], [04/03/ 1993–17/04/1993], and [06/04/1994–20/04/1994]. Let the value of *mininterval* be 10 days. The interval [06/03/1991–12/03/1991] does not satisfy the criterion of *mininterval*. Also let the value of *minsupp* be 0.5. Then {a, c} is not locally frequent in the interval [06/04/1994–20/04/1994]. We shall analyse the pattern {a, c} in the following intervals: [29/03/1990–25/04/1990], [03/03/1992–07/04/1992], and [04/03/1993–17/04/1993]. After superimposition, we require to analyse the interval [03/03–25/04]. We present superimposed intervals in Fig. 9.2.

We present an analysis of the time interval [03/03–25/04] based on the concept of the extended certainty factor. Extended certainty factor of a pattern in an interval provides information of both the certainty factor and *s-info* for a pattern. In Table 9.3, we include an analysis of intervals for determining the yearly pattern {a, c}.

In the above, we have presented an analysis of the time interval [03/03–25/04]. The subintervals [03/03–04/03) and (17/04–25/04] are also shown, but they do not satisfy the *mininterval* criterion. In the experimental results we have not presented such subintervals.

9.4　Mining Calendar-Based Periodic Patterns

Itemsets in transactions could be considered as a basic type of pattern in a database. Many interesting patterns such as association rules (Agrawal et al. 1993), negative association rules (Wu et al. 2004), Boolean expressions induced by itemset (Adhikari and Rao 2007) and conditional patterns (Adhikari and Rao 2008) are based on itemset patterns. Some itemsets are frequent in certain time intervals but may not be frequent throughout the lifespan of the itemsets. In other words, some itemsets may appear in the transactions for a certain time period and then disappear for a long period and then reappear. In view of making a data analysis involving various itemsets, it might be required to extract the itemsets together with the associated time-slots.

9.4.1 Improving Mining Calendar-Based Periodic Patterns

The goal of this chapter is to study the existing algorithm, and to propose an effective algorithm by improving the limitations of the existing algorithm for mining calendar-based periodic patterns. As noted earlier that the concept of certainty factor of an interval does not provide good analysis of overlapped intervals. Therefore, the concept of extended certainty factor has been proposed. In view of designing an effective algorithm, we also need to understand the existing algorithm. Mahanta et al. (2005) have proposed an algorithm for finding all the locally frequent itemsets of size one. While studying the algorithm we have found that some variables contradict their definitions. Authors defined two variables *ptcount* and *ctcount* as follows. The variable *ptcount* is used to count the number of transactions in an interval in which the current item belongs. On the other hand, the variable *ctcount* is used to count the number of transactions in that interval. Therefore, the assignment *ptcount*[k] = *ctcount*[k] in the algorithm, seems to be not appropriate. Also, the variable *icount* is defined as the number of items present in the whole dataset. Therefore, the initialization, *icount* = 1, placed just before starting a new interval seems to be inappropriate. Moreover, the validity of the experimental results is low, since it is based on only one dataset. In view of improving the algorithm further we propose a number of modifications mentioned as follows: (a) The proposed algorithm makes corrections on the existing algorithm using the points noted above. (b) It makes effective data analysis by incorporating extended certainty factor. (c) We propose a hash-based data structure to improve the space efficiency of our algorithm. (d) Also, we have improved the average time complexity of the algorithm. (e) We have completed a comparative analysis with the existing algorithm. (f) In addition, we have improved the validity of the experimental results by conducting experiments on more datasets.

9.4.2 Data Structure

We discuss here the data structure used in the proposed algorithms for mining itemsets along with the time intervals in which they are frequent. We describe the data structure using Example 9.5 given below.

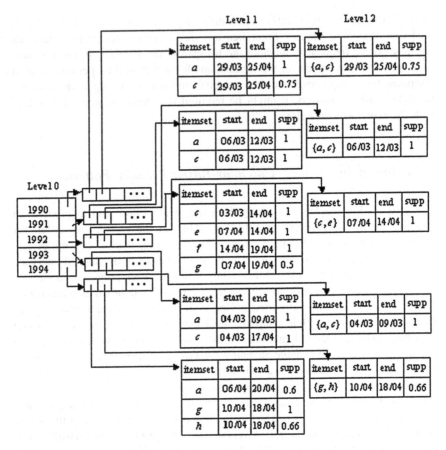

Fig. 9.3 Data structure used in the proposed algorithms

Example 9.5 Consider the database *D* of Example 9.1. Transactions consisting of items *a*, *b*, *c*, *d*, *e*, *f*, *g*, *h*, and *i* occurred in the years from 1990 to 1994. We propose Algorithm 9.1 to mine locally frequent itemsets of size one along with their intervals. The algorithm produces output as shown at level 1 of Fig. 9.3. We assume here that *maxgap*, *mininterval* and *minsupp* assume values of 40 days, 5 days and 0.5, respectively. We are interested in identifying yearly periodic patterns. At the level 0 we have shown all the years that appeared in the transaction. The pointer corresponding to the year 1990 keeps all the locally frequent itemsets of size one, their supports and intervals. All the 5 years are stored in an index table at level 0. After level 0, we keep an array of pointers for every year. The first pointer corresponding to year 1990 points to a table containing interesting itemsets of size one, their intervals and local supports. The second pointer corresponding to year 1990, points to a table containing interesting itemsets of size two, their intervals and local supports, and so on. Here itemsets of size three corresponding to a year do not get reported. Different itemsets, their intervals, and supports are shown in Fig. 9.3. •

9.4.3 A Modified Algorithm

As mentioned in Sect. 9.4.1, we have proposed a number of improvements to the algorithm proposed by Mahanta et al. (2005) for finding locally frequent itemsets of size one. We calculate the support of each item in an interval and store it whenever the item is frequent in that interval. Intervals that satisfy the user-defined constraints *mininterval* and *minsupp* are retained. The modification made seems to be significant from the overall viewpoint of apriori algorithm. We have used a hash-based data structure to improve efficiency of storing and accessing locally frequent itemsets of size one. We explain all the variables and their functions in the following paragraph.

Let *item* be an array of items in *D*. Also let the total number of items be *n*. We use index *level_0* to keep track of different years. It is a two-dimensional array containing 2 columns. First column of *level_0* contains the different years in increasing order. A two-dimensional array *itemset_addr* is used to store the addresses of tables containing itemsets. *itemset_addr*[*row*][*j*] contains the address of the table containing locally frequent itemsets of size *j* for the year *row*. The second column of *level_0* stores addresses of arrays pointing to these tables. Tables at *level_p* store the frequent itemsets of size *p*, $p = 1, 2, 3, \ldots$. Variables *row* and *row_p* are used to index arrays *itemset_addr* and *level_p* respectively, $p = 0, 1, 2, \ldots$. We consider a transaction as a record containing transaction date (*date*) and items purchased. Function *year*() is used to extract year from a given date. *firstseen*[*k*] and *lastseen*[*k*] specify the date when the *k*-th item is seen for the first time and last time in an interval, respectively. Each item in the database is associated with the arrays *itemIntervalFreq* and *nTransInterval*. Cells *itemIntervalFreq*[*k*] and *nTransInterval*[*k*] are used to keep the number of transactions containing item *k* and total number of transactions in a time interval, respectively. Variable *nItemsTrans* is used to keep track of the number of items in the current transaction. The goal of the Algorithm 9.1 is to find all the locally frequent itemsets of size one, their intervals and supports. The algorithm is presented as follows (Adhikari and Rao 2013).

Algorithm 9.1 Mine locally frequent items and their intervals
procedure *MiningFrequentItems* (*D, maxgap, mininterval, minsupp*)

Inputs: *D, maxgap, mininterval, minsupp*
D: database to be mined
minsupp: as defined in Section 1
maxgap, mininterval: as defined in Section 9.3.2
Outputs:
Locally frequent items, their intervals and supports as mentioned in Figure 9.3
01: **let** *nItemsTrans* = 0; *row* = 1; *row_0* = 1; *row_1* = 1;
02: **for** *k* = 1 to *n* **do**
03:      *lastseen*[*k*] = 0; *itemIntervalFreq*[*k*] = 0; *nTransInterval*[*k*] = 0;
04: **end for**
05: read a transaction *t* ∈ *D*;
06: *level_0*[*row_0*][1] = *year*(*t.date*);
07: *level_0*[*row_0*][2] = *itemset_addr*[*row*][1];
08: **while** not end of transaction in *D* **do**
09:      *transLength* = |*t*|;
10:      **if** (*level_0*[*row_0*][1] ≠ *year*(*t.date*)) **then**
11:          **for** *k* = 1 to *n* **do**
12:              **if** (|*lastseen*[*k*] − *firstseen*[*k*]| ≥ *mininterval*) **and**
                 (*itemIntervalFreq*[*k*] / *nTransInterval*[*k*] ≥ *minsupp*) **then**
13:                  store the *k*-th *item*, its *firstseen*, *lastseen* and local support at *level_1*[*row_1*];
14:                  increase *row_1* by 1;
15:              **end if** {12}
16:          **end for** {11}
17:          *row_1* = 1;
18:          increase *row_0* by 1; increase *row* by 1;
19:          *level_0*[*row_0*][1] = *year*(*t.date*); *level_0*[*row_0*][2] = *itemset_addr*[*row*][1];
20:          **for** *k* = 1 to *n* **do**
21:              *lastseen*[*k*] = 0; *itemIntervalFreq*[*k*] = 0; *nTransInterval*[*k*] = 0;
22:          **end for**
23:      **end if** {10}
24:      **for** *k* = 1 to *n* **do**
25:          **if** (*item*[*k*] ∈ *t*) **then**
26:              increase *nItemsTrans* by 1;
27:              **if** (*lastseen*[*k*] = 0) **then**
28:                  initialize both *lastseen*[*k*] and *firstseen*[*k*] by *t.date*;
                 initialize both *itemIntervalFreq*[*k*] and *nTransInterval*[*k*] by 1;
29:              **else if** (| *t.date* − *lastseen*[*k*] | ≤ *maxgap*) **then**
30:                  *lastseen*[*k*] = *t.date*;
31:                  increase *itemIntervalFreq*[*k*] by 1; increase *nTransInterval*[*k*] by 1;
32:              **end if**
33:              **else if** (| *lastseen*[*k*] − *firstseen*[*k*] | ≥ *mininterval*) **and**
                 (*itemIntervalFreq*[*k*] / *nTransInterval*[*k*] ≥ *minsupp*) **then**
34:                  store the *k*-th *item*, its *firstseen*, *lastseen* and local support at *level_1*[*row_1*];
35:                  increase *row_1* by 1;
36:                  initialize both *lastseen*[*k*] and *firstseen*[*k*] by *t.date*;
37:                  initialize both *itemIntervalFreq*[*k*] and *nTransInterval*[*k*] by 0;
38:              **end if** {33}

```
39:       end if {27}
40:     else increase nTransInterval[k] by 1;
41:     end if {25}
42:     if (nItemsTrans = transLength) then exit from for-loop; end if
43:   end for {24}
44:   read a transaction t ∈ D;
45: end while {08}
46: for k = 1 to n do
47:   if (||lastseen[k] − firstseen[k]|| ≥ mininterval) and
        (itemIntervalFreq[k] / nTransInterval[k] ≥ minsupp) then
48:     store the k-th item, its firstseen, lastseen and local support at level_1[row_1];
49:     increase row_1 by 1;
50:   end if {47}
51: end for {46}
52: sort arrays level_1 on non-increasing order on primary key item and
      secondary key start date;
end procedure
```

At line 5 we read the first transaction of database. Afterwards the first row of the index *level_0* is initialized with the first year obtained from the transaction. The pointer field of the first row of *level_0* is initialized by the address of the first row of the table *itemset_addr*. Lines 8–45 are repeated until all the transactions are read. At line 10 we check whether the current transaction belongs to a different year. If it happens so then we close the last interval of different items using lines 11–16. We retain those intervals that satisfy criteria of *mininterval* and *minsupp*. Lines 17–22 assign the necessary initializations for a different year. Lines 25–41 are repeated for each item in the current transaction. Line 27 checks whether the item is first time seen in the transaction and the necessary assignment is done in line 28. Lines 29–32 determine whether the current transaction-date is coming under the current interval by comparing the difference between *t.date* and *lastseen* with *maxgap*. In Lines 33–38 we construct an interval and compute the local support. Line 42 avoids the unnecessary repetition by comparing the transaction length. Line numbers 46–51 close all the last intervals for last year. Line 52 sorts arrays *level_1* on non-increasing order on primary key item and secondary key start date.

The time complexity of the algorithm has been reduced significantly by computing the length of the current transaction (at line number 9) and putting a check at line number 25. Consider a database containing 10,000 items. Let the current transaction be of length 20 and these items are within the first 100 items. Then the *for-loop* at line number 24 need not have to continue for the remaining 9,900 items, but the worst-case complexity of the algorithm remains the same as before.

We now present below an algorithm that makes use of locally frequent itemsets obtained by Algorithm 9.1 and apriori property (Agrawal and Srikant 1994). We use array *level_1* to generate the candidate sets at the second level. Then array *level_2* is used to generate candidate sets at the third level, and so on. We apply pruning using conditions at line number 6 to eliminate some itemsets at the next level. This pruning step ensures that the size of the itemsets at the current level is one more than the size of an itemset at the previous level. Also we apply pruning using user-defined thresholds such as *maxgap*, *mininterval* and *minsupp*. The goal of the Algorithm 9.2 is to find all the locally frequent itemsets of size greater than one, their intervals and supports. The algorithm is presented as follows (Adhikari and Rao 2013).

Algorithm 9.2 Mine locally frequent itemsets at higher level and the associated intervals
procedure *MiningHigherLevelItemsets* (*D, S*)

Inputs: *D, S*
D: database to be mined
S: partially constructed data structure containing locally frequent itemsets of size one
Outputs: locally frequent itemsets at higher levels, the associated intervals and supports as mentioned in Fig. 9.3
01: **let** L_1 = set of elements at *level_1* of *S*; **let** $k = 2$;
02: **while** $L_{k-1} \neq \phi$ **do**
03:    $C_k = \phi$;
04:    **for** each itemset $l_1 \in L_{k-1}$ **do**
05:       **for** each itemset $l_2 \in L_{k-1}$ **do**
06:          **if** $((l_1[1] = l_2[1]) \wedge ... \wedge (l_1[k-2] = l_2[k-2]) \wedge (l_1[k-1] < l_2[k-1]))$ **then**
07:             $c = l_1 \bowtie l_2$; $C_k = C_k \cup c$;
08:          **end if** {06}
09:       **end for** {05}
10:    **end for** {04}
11:    **for** each element $c \in C_k$ **do**
12:       construct intervals for c as mentioned in Algorithm 9.1;
13:       **if** the intervals corresponding to c satisfy *maxgap*, *mininterval* and *minsupp* **then**
14:          add c and the intervals to *level_k* of *S*;
15:       **end if** {13}
16:    **end for** {11}
17:    increase k by 1;
18:    **let** L_k = set of elements at *level_k* of *S*;
19: **end while** {02}
end procedure

Using Algorithms 9.1 and 9.2, one could construct the data structure *S* presented in Fig. 9.3 completely. One can use *S* to determine whether an itemset pattern is fully/partially periodic.

9.5 Experimental Studies

We have carried out several experiments for mining calendar-based periodic patterns on different databases. All the experiments have been implemented on a 2.4 GHz, core i3 processor with 4 GB of memory, running Windows 7 HB, using Visual C++ (version 6.0) software. We present experimental results using *retail* (Frequent itemset mining dataset repository), *BMS-WebView-1* (Frequent itemset mining dataset repository), and *T10I4D100K* (Frequent itemset mining dataset

Table 9.4 Database characteristics

D	NT	ALT	AFI	NI	Size in megabytes
Retail	88,162	11.31	60.54	16,470	3.97
BMS-WebView-1	1,49,639	2.00	44.57	6,714	1.97
T10I4D100K	1,00,000	11.10	1276.12	870	3.83

repository) databases. Since the records in these databases contain only items purchased in transactions, we have attached time-stamps randomly as calendar date for the transactions. The characteristics of the databases are given in Table 9.4.

For the purpose of conducting experiments, each of the databases *retail, BMS-WebView-1* and *T10I4D100K* has been divided into 30 sub-databases, called yearly databases. The characteristics of these databases are given in Table 9.5. Let D, NT, ALT, AFI, and NI be the given database, the number of transactions, average length of a transaction, average frequency of an item, and the number of items, respectively. In Table 9.5 we have shown how the transactions have been time-stamped. The yearly databases obtained from *retail, BMS-WebView-1* and *T10I4D100K* are named as R_i, B_i and T_i respectively, $i = 1, \ldots, 30$. For simplicity, we have kept the number of transactions in each of the yearly databases fixed, except for the last database. We assume that the first and the last transactions occur on 01/01/1961 and 31/12/1990, respectively, and also assume that each year contains 365 days. In our experimental studies we report yearly periodic patterns and the associated periodicities. We also determine the certainty factor and the match ratio of a pattern with respect to overlapped intervals.

In addition to partial periodic patterns, we mine full periodic patterns in the above databases. Itemset patterns of size one and two of *retail* is shown in Tables 9.6 and 9.7 respectively. In *retail* the itemsets {39} and {48} occur in all the 30 years and they are periodic throughout the year. Therefore, these itemsets

Table 9.5 Characteristics of yearly databases

D	NT	Starting date, ending date	Average number of transactions per day
R_1	2920	01/01/1961, 31/12/1961	8
...	...	...	...
R_{29}	2920	01/01/1989, 31/12/1989	8
R_{30}	3482	01/01/1990, 31/12/1990	9.54
B_1	5110	01/01/1961, 31/12/1961	14
...	...	...	...
B_{29}	5110	01/01/1989, 31/12/1989	14
B_{30}	1449	01/01/1990, 31/12/1990	3.97
T_1	3285	01/01/1961, 31/12/1961	9
...	...	...	...
T_{29}	3285	01/01/1989, 31/12/1989	9
T_{30}	4735	01/01/1990, 31/12/1990	12.97

are full periodic in the interval [1/1–31/12]. Itemset {41} is partially periodic, since the match ratio is less than 1. Initially it becomes frequent for 13 years and then it does not get reported, and again it becomes frequent for the last 6 years. The subintervals that do not satisfy the *mininterval* criterion are not shown. We have noticed some peculiarity in the mined patterns. For example, many patterns such as {0} and {1} are frequent throughout a year. Although, it is peculiar but it remains also an artificial phenomenon, since the time-stamps are enforced by us. There are many itemsets such as {16217} are frequent in many years with non-overlapping intervals. In Table 9.6, we present itemsets of size one that are also part of interesting itemsets of size two as shown in Table 9.7. While computing the certainty factor of an itemset we have used lifespan of the itemset. For example, itemset {0} gets reported from 2 years and it becomes frequent in both the years. Therefore its certainty factor is 2/2 = 1.

Interesting itemset patterns of size one and two in *BMS-WebView-1* are shown in Tables 9.8 and 9.9, respectively. Here full periodic patterns are not reported since all the itemsets in *BMS-WebView-1* have the value of the match ratio lower than 1. Therefore, these patterns are partially periodic. Itemset {12355} becomes frequent in 3 years but it has lifespan for 7 years. In this database the items are sparse. Therefore, one requires choosing a smaller *minsupp*. From Table 9.9 one could observe that itemset {33449, 33469} shows periodicity by appearing two times in 6 years and the remaining interesting itemsets are reported for a year only.

Table 9.6 Selected yearly periodic itemsets of size one (for *retail*)

Retail (*minsupp* = 0.25, *mininterval* = 8, *maxgap* = 10)

Itemset	Intervals	Certainty	s-info	Match ratio
{0}	[1/1–31/12]	2/2	0.35–0.66	1.0
{1}	[3/1–31/12]	2/3	0.57–0.66	0.67
{39}	[1/1–31/12]	30/30	0.52–0.63	1.0
{41}	[1/1–22/12]	13/30	0.26–0.32	0.43
{41}	[2/12–30/12]	6/30	0.27–0.32	0.20
{48}	[1/1–31/12]	30/30	0.43–0.53	1.0
{16217}	[1/1–30/5]	1/1	0.87–0.87	1.0
{16217}	[7/9–31/12]	1/1	0.97–0.97	1.0

Table 9.7 Yearly periodic itemsets of size two (for *retail*)

Retail (*minsupp* = 0.25, *mininterval* = 8, *maxgap* = 10)

Itemset	Intervals	Certainty	s-info	Match ratio
{0, 1}	[15/10–31/12]	1/1	0.46	1.0
{39, 41}	[1/1–30/12]	1/1	0.25	1.0
{39, 48}	[1/1–30/12]	30/30	0.28–0.38	1.0
{39, 16217}	[1/1–30/5]	1/1	0.34	1.0
{48, 16217}	[7/9–31/12]	1/1	0.27	1.0

Table 9.8 Yearly periodic itemsets of size one (for *BMS-WebView-1*)

BMS-WebView-1(minsupp = 0.06, mininterval = 7, maxgap = 10)

Itemset	Intervals	Certainty	*s-info*	Match ratio
{10311}	[29/1–6/10]	2/6	0.063–0.86	0.33
{12355}	[21/12–28/12]	3/7	0.060–0.061	0.43
{12559}	[22/4–11/5]	1/2	0.064–0.066	0.5
{33449}	[3/1–26/12]	5/7	0.063–0.08	0.71
{33469}	[3/1–31/3]	5/7	0.067–0.08	0.71

Table 9.9 Yearly periodic itemsets of size two (for *BMS-WebView-1*)

BMS-WebView-1(minsupp = 0.06, mininterval = 7, maxgap = 10)

Itemset	Intervals	Certainty	*s-info*	Match ratio
{10311, 12559}	[30/4–9/4]	1/1	0.06-0.06	1.0
{10311, 33449}	[3/3–11/4]	1/1	0.065	1.0
{33449, 33469}	[15/2–25/3]	2/6	0.061–0.064	0.33

Table 9.10 Selected yearly periodic itemsets of size one (for *T10I4D100K*)

T10I4D100K (minsupp = 0.13, mininterval = 7, maxgap = 10)

Itemset	Interval	*s-info*	Itemset	Interval	*s-info*
{966}	[28/1/1961–19/2/1961]	0.16	{998}	[13/11/1964–22/11/1964	0.17
{966}	[16/3/1961–23/3/1961]	0.17	{998}	[15/12/1964–27/12/1964]	0.16
{966}	[14/12/1961–25/12/1961]	0.16	{998}	[2/9/1965–13/9/1965]	0.14
{966}	[22/3/1964–13/4/1964]	0.15	{998}	[27/11/1966–8/12/1966]	0.14
{966}	[1/11/1975–8/11/1975]	0.16	{998}	[27/11/1973–11/12/1973]	0.13
{966}	[12/4/1981–19/4/1981]	0.17	{998}	[14/12/1983–23/12/1983]	0.17
{966}	[2/12/1988–12/12/1988	0.15	{998}	[15/12/1984–23/12/1984]	0.16

In Table 9.10 we present yearly periodic itemsets of size one for *T10I4D100K* database. In this database patterns with full periodicity are not available, since the intervals corresponding to an item are not overlapped. We have presented examples of such items in the following table. From interval column, one could observe that the itemsets are frequent for the short intervals, but do not appear at the same time for all the years. For example, itemset {966} appears in three intervals in 1961, but it does not show any periodicity since the intervals are not overlapped. It is interesting to note that the itemset {966} appears at the beginning, both in 1st and 2nd months, of the year, then at the middle of the year i.e., for the 3rd and 4th months, and finally at the end of the year (11th and 12th month). This is also true for itemset {998}. Interesting itemset patterns of size two are not reported in this database.

An itemset that satisfies *minsupp, mininterval* criteria are reported. Also, a locally frequent itemset in two intervals for a particular year is also reported in the intervals, provided the intervals satisfy *maxgap* criterion. The number of interesting intervals could increase by lowering the thresholds. In the following paragraphs we have presented a study of these aspects.

9.5.1 Selection of Mininterval and Maxgap

The selection of *mininterval* and *maxgap* might be crucial since the process of data mining would depend on factors like seasonality, type of application and the data source. Some items are used for a particular season; while others are purchased throughout the year. When the items are purchased throughout the year, the choices of *mininterval* and *maxgap* do not have much significance in mining yearly patterns. This observation seems to be valid for the items in *retail* and *BMS-WebView-1*. But the items in *T10I4D100K* are frequent in smaller intervals and therefore, *mininterval* and *maxgap* might have an impact on data mining. On the other hand, the requirement of an organization might determine an important parameter for mining calendar-based patterns. The distribution of items in databases also matters in selecting the right values of *mininterval* and *maxgap*. For a sparse database *maxgap* could be longer, and it could be even longer than *mininterval* provided *minsupp* remains low.

9.5.1.1 Mininterval

In the following experiments we would like to analyse the effect of *mininterval* for given *maxgap* and *minsupp*. We observe in Figs. 9.4, 9.5 and 9.6, the number of intervals decreases as *mininterval* increases. An itemset might be frequent in many intervals. The number of itemsets frequent in an interval decreases as the length of *mininterval* increases. Although the above observation is true in general, but the type of the graphs might differ from one data source to another. In *retail* many itemsets are locally frequent for longer period of time. In Fig. 9.4 we observe that there exists nearly 110 intervals for *mininterval* of 29 days. Whereas in

Fig. 9.4 *Retail* (*minsupp* = 0.25, *maxgap* = 7)

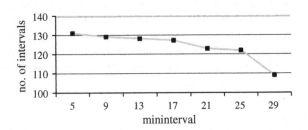

Fig. 9.5 *BMS-WebView-1* (*minsupp* = 0.06, *maxga*p = 7)

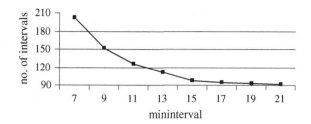

Fig. 9.6 *T10I4D100K* (*minsupp* = 0.13, *maxga*p = 7)

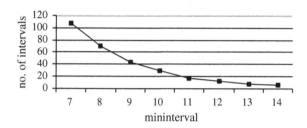

BMS-WebView-1 and *T10I4D100K*, the itemsets are frequent for shorter duration. As a result, the number of intervals reduces significantly when *mininterval* remains small. Thus the choice of *mininterval* is an important issue.

9.5.1.2 Maxgap

In view of analyzing *maxgap* parameter, we present graphs of the number of intervals versus *maxgap* at given *minsupp* and *mininterval* in Figs. 9.7, 9.8, and 9.9. The graphs show that the number of intervals decreases as *maxgap* increases.

Fig. 9.7 *Retail* (*minsupp* = 0.25, *mininterval* = 10)

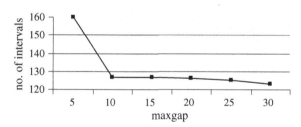

Fig. 9.8 *BMS-WebView-1* (*minsupp* = 0.06, *mininterval* = 7)

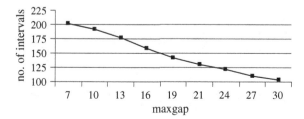

Fig. 9.9 *T10I4D100K*
(*minsupp* = 0.13,
mininterval = 7)

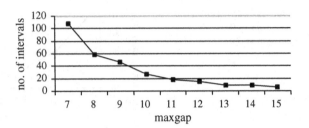

In *retail* the number of intervals decreases rapidly when *maxgap* varies from 5 to 10. Afterwards the change is not so significant. In *BMS-WebView-1* the decrement takes place almost at a uniform rate. Unlike *retail* and *BMS-WebView-1*, the number of intervals decreases faster at the smaller values of *maxgap* in *T10I4D100K* dataset.

9.5.2 Selection of Minsupp

The number of intervals and minimum support are inversely related to given *maxgap* and *mininterval*. We observe this phenomenon in Figs. 9.10, 9.11, and 9.12. When the value of *maxgap* is smaller, the number of intervals reported is quite large. Initially the number of intervals decreased significantly with small decrement of *minsupp*. Later the decrement of the number of intervals is not so significant.

Fig. 9.10 *Retail*
(*mininterval* = 10,
maxgap = 12)

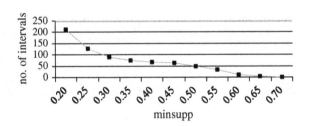

Fig. 9.11 *BMS-WebView-1*
(*mininterval* = 7,
maxgap = 10)

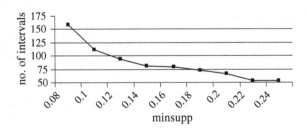

Fig. 9.12 *T10I4D100K* (*mininterval* = 7, *maxgap* = 9)

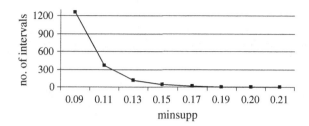

9.5.3 Performance Analysis

In this section, we present the performance of our algorithm and compare it with the performance of the existing algorithm for mining calendar-based periodic patterns. To quantify the performance, two experiments have been conducted. In the first experiment, we have measured the scalability of the two algorithms with respect to different database sizes. In the second experiment, we have measured the scalability of the two algorithms with respect to different support thresholds. In Figs. 9.13, 9.14, and 9.15, we have shown the relationship between the database size and execution time for mining periodic patterns. We observed that the number of patterns increases as the number of transactions increases. Thus, the execution time increases with the increase of the database size. Initially both the algorithms take almost equal amount of time. In Figs. 9.13 and 9.14 we observe that execution time for mining 88,162 transactions of *retail* and 1,49,639 transactions of

Fig. 9.13 Execution time versus size of database at *minsupp* = 0.25, *mininterval* = 8, *maxgap* = 10 (*retail*)

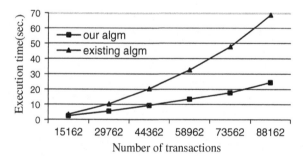

Fig. 9.14 Execution time versus size of database at *minsupp* = 0.1, *mininterval* = 7, *maxgap* = 10 (*BMS-WebView-1*)

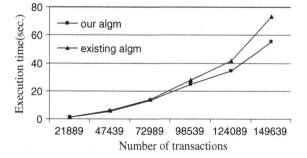

Fig. 9.15 Execution time versus size of database at *minsupp* = 0.13, *mininterval* = 7, *maxgap* = 10 (*T10I4D100K*)

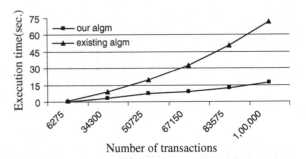

Number of transactions

Fig. 9.16 Execution time versus *minsupp* (*mininterval* = 8, *maxgap* = 10) for *retail*

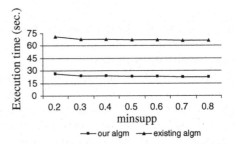

BMS-WebView-1 take nearly equal amount of time. The reason is that the average length of transactions in *retail* is more than that of *BMS-WebView-1*. Therefore, the execution time is not only dependent on the size of the database, but also depends on the factors such as *ALT* and *NI*. The experimental results in Figs. 9.13, 9.14, and 9.15 show that our algorithm performs better than the existing algorithm.

In Figs. 9.16, 9.17, and 9.18, we have presented another comparison by considering *minsupp* threshold. When the minimum support increases, the number of frequent itemsets decreases and subsequently the execution time also decreases. The experimental results have shown that the execution time of both the algorithms decreases slowly when the support threshold increases, and our algorithm takes less time than the existing algorithm.

Fig. 9.17 Execution time versus *minsupp* (*mininterval* = 7, *maxgap* = 10) for *BMS-WebView-1*

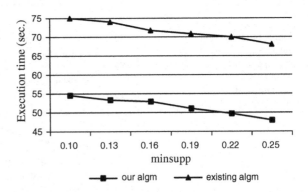

Fig. 9.18 Execution time versus *minsupp* (*mininterval* = 7, *maxga*p = 10) for *T10I4D100K*

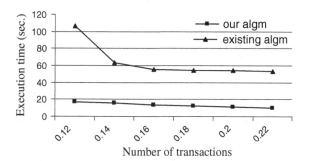

9.6 Conclusions

In this chapter, we have proposed modifications to the existing algorithm for mining locally frequent itemsets along with the set of intervals and associated supports. We have also extended the concept of the certainty factor for a detailed analysis of overlapped intervals. We have proposed a number of improvements of the existing algorithm for finding calendar-based periodic patterns. For managing locally frequent itemsets effectively we have introduced a hash-based data structure. We have presented an extensive data analysis by involving constraints such as *mininterval*, *minsupp* and *maxga*p. In addition we have compared our algorithm with the existing algorithm. Experimental results show that the proposed algorithm runs faster than the existing algorithm. Experimental results also report that whether a periodic pattern is full or partial. The proposed algorithm can also be used to extract yearly, monthly, weekly and daily calendar-based patterns.

References

Adhikari A, Rao PR (2007) A framework for mining arbitrary Boolean expressions induced by frequent itemsets. In: Proceedings of Indian international conference on artificial intelligence, pp 5–23

Adhikari A, Rao PR (2008) Mining conditional patterns in a database. Pattern Recogn Lett 29(10):1515–1523

Adhikari J, Rao PR (2013) Identifying calendar-based periodic patterns. In: Ramanna S, Jain LC, Howlett RJ (ed) Emerging paradigms in machine learning. Springer, Berlin, pp 329–357

Agrawal R, Imielinski T, Swami A (1993) Mining association rules between sets of items in large databases. In: Proceedings of the ACM SIGMOD conference management of data, pp 207–216

Agrawal R, Srikant R (1994) Fast algorithms for mining association rules. In: Proceedings of 20th very large databases (VLDB) conference, pp 487–499

Agrawal R, Srikant R (1995) Mining sequential patterns. In: Proceedings of international conference on data engineering (ICDE), pp 3–14

Ale JM, Rossi GH (2000) An approach to discovering temporal association rules. In: Proceedings of ACM symposium on applied computing, pp 294–300

Allen JF (1983) Maintaining knowledge about temporal intervals. Commun ACM 26(11):832–843

Baruah HK (1999) Set superimposition and its application to the theory of fuzzy sets. J Assam Sci Soc 10(1–2):25–31

Bettini C, Jajodia S, Wang SX (2000) Time granularities in databases data mining and temporal reasoning. Springer, Berlin

Frequent itemset mining dataset repository. http://fimi.cs.helsinki.fi/data

Han J, Dong G, Yin Y (1999) Efficient mining on partial periodic patterns in time series database. In: Proceedings of fifteenth international conference on data engineering, pp 106–115

Hong TP, Wu YY, Wang SL (2009) An effective mining approach for up-to-date patterns. Expert Syst Appl 36:9747–9752

Kempe S, Hipp J, Lanquillon C, Kruse R (2008) Mining frequent temporal patterns in interval sequences. Int J Uncertainty Fuzziness Knowl Based Syst 16(5):645–661

Lee JW, Lee YJ, Kim HK, Hwang BH, Ryu KH (2002) Discovering temporal relation rules mining from interval data. In: Proceedings of the 1st EuroAsian conference on advance in information and communication technology, pp 57–66

Lee YJ, Lee JW, Chai D, Hwang B, Ryu KH (2009) Mining temporal interval relational rules from temporal data. J Syst Softw 82(1):155–167

Lee G, Yang W, Lee JM (2006) A parallel algorithm for mining multiple partial periodic patterns. Inf Sci 176(24):3591–3609

Li D, Deogun JS (2005) Discovering partial periodic sequential association rules with time lag in multiple sequences for prediction. Foundations of Intelligent Systems LNCS, vol 3488, pp 1–24

Li Y, Ning P, Wang XS, Jajodia S (2003) Discovering calendar-based temporal association rules. Data Knowl Eng 44(2):193–218

Mahanta AK, Mazarbhuiya FA, Baruah HK (2005) Finding locally and periodically frequent sets and periodic association rules. Pattern recognition and machine intelligence LNCS, vol 3776, pp 576–582

Mahanta AK, Mazarbhuiya FA, Baruah HK (2008) Finding calendar-based periodic patterns. Pattern Recogn Lett 29(9):1274–1284

Ozden B, Ramaswamy S, Silberschatz A (1998) Cyclic association rules. In: Proceedings of 14th international conference on data engineering, pp 412–421

Roddick JF, Spiliopoulou M (2002) A survey of temporal knowledge discovery paradigms and methods. IEEE Trans Knowl Data Eng 14(4):750–767

Verma K, Vyas OP, Vyas R (2005) Temporal approach to association rule mining using T-tree and P-tree. Machine learning and data mining in pattern recognition LNCS, vol 3587, pp 651–659

Wu X, Zhang C, Zhang S (2004) Efficient mining of both positive and negative association rules. ACM Trans Inf Syst 22(3):381–405

Zimbrao G, de Souza JM, de Almeida VT, de Silva WA (2002) An algorithm to discover calendar-based temporal association rules with item's lifespan restriction. In: Proceedings of the eighth ACM SIGKDD international conference on knowledge discovery and data mining

Chapter 10
Measuring Influence of an Item in Time-Stamped Databases

Influence of items on some other items might not be the same as the association between these sets of items. Many tasks of data analysis are based on expressing influence of items on other items. In this chapter, we introduce the notion of an overall influence of a set of items on another set of items. We also propose an extension to the notion of overall association between two items in a database. Using this notion, we have designed two algorithms of influence analysis involving specific items in a database. As the number of databases increases on a yearly basis, we have adopted incremental approach to these algorithms. Experimental results are reported for both synthetic and real-world databases.

10.1 Introduction

Every time a customer interacts with business, we have an opportunity to gain strategic knowledge. Transactional data contain a wealth of information about customers and their purchase patterns. In fact, these data could be one of the most valuable assets, when used wisely. This has been recognized a long time ago by many large organizations such as supermarkets, insurance companies, healthcare organizations, telecommunications, and banks. These organizations have spent significant resources for collecting and analyzing transactional data. Many applications are based on inherent knowledge present in a database. Such applications could be dealt with mining databases (Han et al. 2000; Agrawal and Srikant 1994; Savasere et al. 1995; Gary and Petersen 2000). As a database changes over time, the inherent knowledge also changes. Therefore, in the competitive market, knowledge-based decisions are more appropriate. Data mining algorithms are effective tools to support making such decisions. Data mining algorithms often extract different patterns from a database. Some examples of patterns in a database are frequent item sets (Agrawal et al. 1993), association rules (Agrawal et al. 1993), negative association rules (Wu et al. 2004), Boolean expressions induced by itemset (Adhikari and Rao 2007) and conditional patterns (Adhikari and Rao 2008a). Nevertheless, there are some applications for which association-based analysis might be inappropriate.

A. Adhikari et al., *Data Analysis and Pattern Recognition in Multiple Databases*,
Intelligent Systems Reference Library 61, DOI: 10.1007/978-3-319-03410-2_10,
© Springer International Publishing Switzerland 2014

For example, an organization might deal with a large number of items with its customers. The company might be interested in knowing how the purchase of a particular item affects the purchase of some other item. In this chapter, we study such influences based on transactional time-stamped database.

Many companies transact a large number of products (items) with their customers. It might be required to perform data analyses involving different items. Such analyses might originate from different applications. One such analysis is identifying stable items (Adhikari et al. 2009) in databases over time. It could be useful in devising strategies for a company. Little work has been reported on data analyses realized over time. In this chapter, we present another application involving different items in a database formed over time.

Consider a company that collects a huge amount of transactional data on a yearly basis. Let DT_i be the database corresponding to the ith year, $i = 1, 2, ..., k$. Each of these databases corresponds to a specific period of time. Thus, one could call these time databases. Each time database is mined using a traditional data mining technique (Adhikari et al. 2011). In this application, we will deal with itemsets in a database. An itemset is a set of items in the database. Let I be the set of all items in the time databases. Each itemset X in a database D is associated with a statistical measure, called *support* (Agrawal et al. 1993), denoted by $supp(X, D)$. The support of an itemset is defined as the fraction of transactions containing the itemset.

Solutions to many problems are based on the study of relationships among variables. We will see later that the study of influence of a set of variables on another set of variables might not be the same as studying the association between these two sets of variables. Association analysis among variables has been studied well (Agrawal et al. 1993; Adhikari et al. 2011; Brin et al. 1997; Shapiro 1991; Adhikari and Rao 2008b, c). In the context of studying association among variables using association rules one could conclude that the confidence of the association rule gives positive influence of antecedent on the consequent of the association rule. Such positive influences might not be sufficient for many data analyses.

Consider an established company possessing data being collected over 50 consecutive years. Generally, the sales of a product vary from one season to another. Also, a season re-appears on a yearly basis. Thus, we divide the entire database into a sequence of yearly databases. In this context, a yearly database could be considered as a time database. In this study, we estimate the influence of item x on y, for $x, y \in I$ where I is the set of all items in database D. In Sect. 10.3, we define the concept of influence of an itemset on another itemset.

An itemset could be viewed as a basic type of pattern in a database. Different types of pattern in a database could be derived from itemset patterns. For example, frequent itemset, association rule, negative association rule, Boolean expression induced by itemset and conditional pattern are examples of derived patterns in a database. Few applications have been reported on analysis of patterns over time. In this chapter, we wish to study the influence of an item on a specific item/a set of specific items in a database.

Most of the association analyses are based on a positive association between variables. Such positive association gives rise to positive influence of variables on other variables. Most of the real databases are large and sparse. In such cases an association analysis using positive influence might not be appropriate, if the overall influence of former variable on latter variable becomes negative. Thus, the concept of overall influence needs to be introduced.

The chapter is organized as follows. In Sect. 10.2, we extend the notion of overall association between two items in a database. In Sect. 10.3, we introduce the notion of overall influence of an itemset on another itemset in a database. We study various properties of proposed measures. Also, we introduce the notion of overall influence of an item on a set of specific items in a database. In addition, we discuss the motivation of the proposed problem in this section. We state our problem in Sect. 10.4. We discuss work related to the proposed problem in Sect. 10.5. In Sect. 10.6, we design an algorithm to measure the overall influence of an item on another item (incrementally). In addition, we design another algorithm of overall influence of an item on a set of specific items (incrementally). Experimental results are provided in Sect. 10.7. We conclude the chapter in Sect. 10.8.

10.2 Association Between Two Itemsets

Adhikari et al. (2011) have proposed a measure denoted by OA, for computing an overall association between two items in a market basket data. Using positive association PA between two items (Adhikari et al. 2011), one could extend positive association between two itemsets in a database as follows:

$$PA(X, Y, D) = \frac{\# \text{ transaction containing both } X \text{ and } Y, D}{\# \text{ transaction containing at least one of } X \text{ and } Y, D},$$

where X and Y are itemsets in database D and "#P, D" is the number of transactions in D that satisfy the predicate P.

Similarly, negative association NA between two items (Adhikari et al. 2011) could be extended as follows:

$$NA(X, Y, D) = \frac{\# \text{ transaction containing exactly one of } X \text{ and } Y, D}{\# \text{ transaction containing at least one of } X \text{ and } Y, D},$$

where X and Y are itemsets in database D.

Using PA and NA, OA between two itemsets X and Y in database D could be defined as follows:

$$OA(X, Y, D) = PA(X, Y, D) - NA(X, Y, D) \qquad (10.1)$$

If $OA(X, Y, D)$ is positive, negative or zero then all the items in X together and all the items in Y together are positively, negatively or independently associated

Table 10.1 Supports of itemsets in D_1

Itemset($\{X\}$)	$\{a, b\}$	$\{c, d\}$	$\{a, c\}$	$\{b, d\}$	$\{d, e\}$	$\{e, g\}$
$supp(\{X\}, D_1)$	4/9	4/9	3/9	3/9	4/9	2/9

Table 10.2 Overall association between two itemsets in D_1

Itemset($\{X, Y\}$)	$\{\{a, b\}, \{c, d\}\}$	$\{\{a, c\}, \{b, d\}\}$	$\{\{c\}, \{d, e\}\}$
$OA(X, Y, D_1)$	1/5	1	$-3/7$

in D, respectively. We illustrate different types of association using the following example.

Example 10.1 Let database D_1 contain the following transactions: $\{a, d, e\}$, $\{a, b, c, d, g\}$, $\{a, b, e, g\}$, $\{b, c, g\}$, $\{d, e, g\}$, $\{b, e, f\}$, $\{c, d, e, f\}$, $\{a, b, c, d, f, g\}$, and $\{a, b, c, d, e\}$. We find here overall association between itemsets X, and Y, for some X, Y in D_1. In Table 10.1, supports of some itemsets are given.

Here $PA(\{a, b\}, \{c, d\}, D_1) = 3/5$ and $NA(\{a, b\}, \{c, d\}, D_1) = 2/5$. Therefore, $OA(\{a, b\}, \{c, d\}, D_1) = 1/5$. In Table 10.2, we show overall associations.

Considering these results, we observe that the OA values between $\{a, b\}$ and $\{c, d\}$ as well as $\{a, c\}$ and $\{b, d\}$ are positive. But the OA value between $\{c\}$ and $\{d, e\}$ is negative. •

10.3 Concept of Influence

Let X and Y be two itemsets in database D. We wish to find influence of X on Y in D. In the previous section, we have proposed overall association between two itemsets. The influence of X on Y seems to be different from the overall association between X and Y.

Let $X = \{x_1, x_2, \ldots x_p\}$ and $Y = \{y_1, y_2, \ldots y_q\}$ be two itemsets in database D. The influence of X on Y could be judged based on the following events: (1) whether a customer purchases all the items of Y when he/she purchases all the items of X, and (2) whether a customer purchases all the items of Y when they do not purchase all the items of X. Such behaviors could be modeled using supports of $X \cap Y$ and $\neg X \cap Y$. The expression $supp(X \cap Y, D)/supp(X, D)$ measures the strength of positive association of X on Y. The expression $supp(\neg X \cap Y, D)/supp(\neg X, D)$ measures the strength of negative association of X on Y. Thus, the expressions $supp(X \cap Y, D)/supp(X, D)$ and $supp(\neg X \cap Y, D)/supp(\neg X, D)$ could be important in measuring an overall influence of X on Y.

10.3.1 Influence of an Itemset on Another Itemset

Let X and Y be the two itemsets in database D. The interestingness of an association rule r_1: $X \to Y$ could be expressed by its support and confidence (conf) measures (Agrawal et al. 1993). These measures are defined as follows. $supp(r_1, D) = supp(X \cap Y, D)$, and $conf(r_1, D) = supp(X \cap Y, D)/supp(X, D)$. The measure $conf(r_1, D)$ could be interpreted as the fraction of transactions containing itemset Y among the transactions containing X in D. In other words, $conf(r_1, D)$ could be viewed as the *positive influence* (PI) of X on Y. Let us consider the negative association rule r_2: $\neg X \to Y$. Confidence of r_2 in D could be viewed as fractions of transactions containing Y among the transactions containing $\neg X$. In other words, confidence of r_2 in D could be viewed as *negative influence* (NI) of X on Y. Similarly to the overall association defined in (10.1), one could define *overall influence* (OI) of X on Y in a database as follows:

Definition 10.1 Let X and Y be two itemsets in database D such that $X \cap Y = \phi$. Then overall influence of X on Y in D is defined as follows (Adhikari and Rao 2010):

$$OI(X, Y, D) = supp(X \cap Y, D)/supp(X, D) - supp(\neg X \cap Y, D)/supp(\neg X, D) \bullet \quad (10.2)$$

$OI(X, Y, D)$ represents the difference of the influence on Y when X is present in a transaction and the influence on Y when X is not present in the transaction. Let γ be user-defined level of interestingness. Then $OI(X, Y, D)$ is *interesting* if $OI(X, Y, D) \geq \gamma$.

If $OI(X, Y, D) > 0$ then the itemset X has positive influence on itemset Y in D. In other words, all the items in X together help promoting itemset Y in D. If $OI(X, Y, D) < 0$ then X has negative influence on Y in D. In other words, all the items in X in D together do not help promoting together all the items in Y. If $OI(X, Y, D) = 0$ then X has no influence on Y in D. In Example 10.2, we illustrate the concept of overall influence.

Example 10.2 We continue our discussion we started in Example 10.1. We have $PI(\{a, b\}, \{c, d\}, D_1) = 3/4$, $NI(\{a, b\}, \{c, d\}, D_1) = 1/5$, and $OI(\{a, b\}, \{c, d\}, D_1) = 11/20$. We observe that $PI(\{a, b\}, \{c, d\}, D_1)$ is more than $PA(\{a, b\}, \{c, d\}, D_1)$. Also, $NA(\{a, b\}, \{c, d\}, D_1)$ is more than $NI(\{a, b\}, \{c, d\}, D_1)$. So, $OI(\{a, b\}, \{c, d\}, D_1)$ is more than $OA(\{a, b\}, \{c, d\}, D_1)$. In similar to overall association, overall influence could be negative as well. Let $X = \{c\}$ and $Y = \{d, e\}$. $PI(X, Y, D_1) = 2/5$, $NI(X, Y, D_1) = 1/2$, and $OI(X, Y, D_1) = -1/10$. Thus, overall influence between two itemsets could be negative as well as positive. $\bullet$

In most of the cases, the overall influence between two itemsets in a large database is negative. In real-world databases, it might be possible that the overall influence between the two itemsets is positive. In Example 10.3, we consider some special cases to illustrate the measure of overall influence.

Example 10.3 Let database D_2 contain the following transactions: $\{a, b, e\}$, $\{a, e, g\}$, $\{b, e, g\}$, $\{a, b, d, e, g\}$, $\{b, d, e, g\}$ and $\{c, e, g\}$. We compute overall influence of an itemset X on another itemset Y considering various cases.

Case 1 $supp(X, D_2) > supp(Y, D_2)$
   Let $X = \{e, g\}$, $Y = \{a, b\}$. $supp(X, D_2) = 5/6$, $supp(Y, D_2) = 2/6$ and $supp(X \cap Y, D_2) = 1/6$. We obtain $OI(X, Y, D_2) = -0.8$.

Case 2 $supp(X, D_2) < supp(Y, D_2)$
   Let $X = \{a, b\}$, $Y = \{e, g\}$. $supp(X, D_2) = 2/6$, $supp(Y, D_2) = 5/6$ and $supp(X \cap Y, D_2) = 1/6$. We have $OI(X, Y, D_2) = -0.5$.

   Though the values of overall influence are negative for the above cases, the influence might turn out to be positive for some databases. Let us consider another database $D_3 = \{\{a, b, c, d, g\}, \{b, c, g\}, \{c, d, g\}, \{a, b, c, d, e\}, \{b, c, e, g\}, \{a, b, c, d, e, g\}\}$.

Case 1 $supp(X, D_3) > supp(Y, D_3)$
   Let $X = \{c, d\}$, $Y = \{a, b\}$. $supp(X, D_3) = 4/6$, $supp(Y, D_3) = 3/6$ and $supp(X \cap Y, D_3) = 3/6$. We have $OI(X, Y, D_3) = 0.5$.

Case 2 $supp(X, D_3) < supp(Y, D_3)$. Let $X = \{a, b\}$, $Y = \{c, d\}$. $supp(X, D_3) = 3/6$, $supp(Y, D_3) = 4/6$ and $supp(X \cap Y, D_3) = 3/6$. Here we have $OI(X, Y, D_3) = 0.667$.

10.3.2 Properties of Influence Measures

For the purpose of computing influence of an itemset on another itemset, one needs to express *OI* in terms of supports of relevant itemsets. From (10.2), we get *OI* as follows:

$$OI(X, Y, D) = supp(X \cap Y, D)/supp(X, D) \\ - (supp(Y, D) - supp(X \cap Y, D))/(1 - supp(X, D))$$

Finally, we compute *OI* as follows:

$$OI(X, Y, D) = \frac{supp(X \cap Y, D) - supp(X, D) \times supp(Y, D)}{supp(X, D)[1 - supp(X, D)]}, \text{ if } supp(X, D) \neq 1$$

and $supp(Y, D) \neq 1$

$$OI(X, Y, D) = 0, \quad \text{otherwise} \qquad (10.3)$$

   From the above formula one could observe that if support of itemset X in D is 1 then influence of other itemsets on X will be zero. On the other hand, if $supp(Y, D) = 1$ then $supp(X \cap Y, D) = supp(X, D)$ and $supp(X, D) \times supp(Y, D) = supp(X, D)$. Therefore, the numerator of formula (10.3) will result to be zero and overall influence becomes zero. In the following lemma, we mention some properties of PI and NI. $OI(X, X, D) = 1$ at $X = Y$. Thus, $OI(X, X, D)$ at $X = Y$ could be termed as *trivial influence*.

Lemma 10.1 *For itemsets X, Y in D, the following properties are satisfied*: (1) $0 \leq PI(X, Y, D) \leq 1$, (2) $0 \leq NI(X, Y, D) \leq 1$, (3) $-1 \leq OI(X, Y, D) \leq 1$.

Lemma 10.2 $OI(X, Y, D) = \frac{supp(Y)[Corr(X,Y,D)-1]}{1-supp(X)}$, *where Corr(X, Y, D) is the correlation coefficient between itemsets X and Y in database D. If Corr(X, Y, D) = 1 then X and Y are independent in database D. In other words, if OI(X, Y, D) = 0 then X and Y are independent in D. If Corr(X, Y, D) < 1 then X and Y are negatively correlated in database D. In other words, if OI(X, Y, D) < 0 then X and Y are negatively correlated. If Corr(X, Y, D) > 1 then X and Y are positively correlated in database D. If OI(X, Y, D) > 0 then X and Y are positively correlated.*

10.3.3 Influence of an Item on a Set of Specific Items

Let $I = \{i_1, i_2, ..., i_m\}$ be the set of items in database D. Also, let $SI = \{s_1, s_2, ..., s_p\}$ be the set of specific items in database D. We would like to analyze the overall influence of each item on SI. The influence of an item on SI could be computed based on $OI(i_j, s_k, D)$, for $j = 1, 2,..., m$ and $k = 1, 2,..., p$. We say that the influence of i_j on s_k is interesting if $OI(i_j, s_k, D) \geq \gamma$, for $j = 1, 2,..., m$ and $k = 1, 2,..., p$. The value of γ depends on the level of data analysis to be performed. If the data analysis is performed in-depth then the value of γ is expected to be low. Also, the value of γ is dependent on the data to be analyzed. Normally, when the data are sparse the user needs to provide a low value of γ. On the other hand, γ could be given a reasonably high value for analyzing dense data. The procedure of determining influence of an item on a set of specific items could be explained using the following steps.

(1) Generate influence matrix (IM) of order $p \times m$ using $OI(i_j, s_k, D)$, for $j = 1, 2,..., p$ and $k = 1, 2,..., m$. (2) An influence is counted when it is interesting. (3) For each item, count the number of interesting influences on each of the specific items. (4) The items in database D are sorted based on primary key as the number of interesting influences on the specific items, and secondary key as the support of an item. We explain steps (1)–(4) using Example 10.4.

Example 10.4 Consider the database D_1 given in Example 10.1. Let $I = \{a, b, c, d, e, f, g\}$ and $SI = \{a, c, d\}$ (Table 10.3).

Table 10.3 Supports of each items in D_1

Items (x)	a	b	c	d	e	f	g
$supp(\{x\}, D_1)$	5/9	6/9	5/9	6/9	6/9	3/9	5/9

In this case, the influence matrix of size 3×7 is given below.

$$
IM = \begin{bmatrix}
item & a & b & c & d & e & f & g \\
a & 1 & 0.333 & 0.100 & 0.333 & -0.167 & -0.333 & 0.350 \\
c & 0.100 & 0.333 & 1 & 0.333 & -0.667 & 0.167 & 0.350 \\
d & 0.300 & -0.500 & 0.300 & 1 & 0 & 0 & -0.150
\end{bmatrix}
$$

Let γ be set to 0.2. Also let $x(\eta)$ denote η number of interesting influences of item x on different specific items. The numbers of interesting influences of different items in D_1 are given as follows. $a(2)$, $b(2)$, $c(2)$, $d(3)$, $e(0)$, $f(0)$, $g(2)$. The items being sorted using step (4) are given as follows: $d(3)$, $b(2)$, $a(2)$, $c(2)$, $g(2)$, $e(0)$, $f(0)$. Given the set of specific items $\{a, c, d\}$, one could conclude that the item d has the maximum and the item f has a minimum influence on the specific items.

10.3.4 Motivation

The concept of influence might not be new in the literature of data mining. For example, $conf(X \rightarrow Y, D)$ refers to positive influence of X on Y. In other words, it implies how likely a customer purchases the items of Y when he has already purchased all the items of X. In addition, the concept of negative influence is present in the literature on data mining. $conf(\neg X \rightarrow Y, D)$ refers to the amount of negative influence of items of X in purchasing the items of Y. In other words, it implies how likely a customer purchases the items of Y when the customer has not purchased all the items of X. In many data analyses it might be required to consider the overall influence of a set of items on another set of items. Our work introduces the notion of overall influence that could be useful in dealing with many real life problems. In the following paragraph, we justify that an existing measure might not be appropriate to study the overall influence of an itemset on another itemset.

The analysis of relationships among variables is a fundamental task being at the heart of many data mining problems. For example, metrics such as support, confidence, lift, correlation, and collective strength have been used extensively to evaluate the interestingness of association patterns. These metrics are defined in terms of the frequency counts tabulated in a 2×2 contingency table as shown in Table 10.4. To illustrate this, let us consider ten example contingency tables, E_1–E_{10}, given in Table 10.5. Tan et al. (2003) presented an overview of twenty-one interestingness measures proposed in the statistics, machine learning, and data mining literature.

In the following discussion, we observe why these measures fail to compute overall influence of an itemset on another itemset. In Examples 10.2 and 10.3, we

Table 10.4 A 2×2 contingency table for variables x and y

	Y	$\neg Y$	Total
X	f_{11}	f_{10}	f_{1+}
$\neg X$	f_{01}	f_{00}	f_{0+}
Total	$f_{.+1}$	$f_{.+0}$	N

Table 10.5 Examples of contingency tables

Example	f_{11}	f_{10}	f_{01}	f_{00}
E1	8,123	83	424	1,370
E2	8,330	2	622	1,046
E3	9,481	94	127	298
E4	3,954	3,080	5	2,961
E5	2,886	1,363	1,320	4,431
E6	1,500	2,000	500	6,000
E7	4,000	2,000	1,000	3,000
E8	4,000	2,000	2,000	2,000
E9	1,720	7,121	5	1,154
E10	61	2,483	4	7,452

have observed that the overall influence of an itemset on another itemset could be positive as well as negative. Thus, overall influence of an itemset on another itemset in a database lies in $[-1, 1]$. In a large database, where items are sparsely distributed over the transactions might result in negative overall influence of an itemset on another itemset. Based on these observations, one could consider the following five out of twenty-one interestingness measures since overall influence of an itemset on another itemset lies in $[-1, 1]$. These measures are presented in Table 10.6.

Based on each formula present in Table 10.6, the above contingency tables have been ranked as shown in Table 10.7. A contingency table that gives the maximum value is ranked as number 1 based on the interestingness measure. For example, contingency tables $E1$ and $E2$ give maximum and the second maximum values based on ϕ.

Also, we rank the contingency tables based on the concept of overall influence explained in Example 10.1. In Table 10.8, we present the ranking of contingency tables when using the overall influence given by (10.3).

None of the five measures ranks contingency tables like the ranks given in Table 10.7. Thus, none of the above five measures serves as a measure of overall influence between two itemsets.

Table 10.6 Relevant interestingness measures for association patterns

Symbol	Measure	Formula
ϕ	ϕ-coefficient	$\dfrac{P(\{x\}\cup\{y\}) - P(\{x\})\times P(\{y\})}{\sqrt{P(\{x\})\times P(\{y\})\times(1-P(\{x\})\times(1-P(\{y\})))}}$
Q	Yule's Q	$\dfrac{P(\{x\}\cup\{y\})\times P(\neg(\{x\}\cap\{y\}))-P(\{x\}\cup\neg\{y\})\times P(\neg\{x\}\cup\{y\})}{P(\{x\}\cup\{y\})\times P(\neg(\{x\}\cap\{y\}))-P(\{x\}\cup\neg\{y\})\times P(\neg\{x\}\cup\{y\})}$
Y	Yule's Y	$\dfrac{\sqrt{P(\{x\}\cup\{y\})\times P(\neg(\{x\}\cap\{y\}))}-\sqrt{P(\{x\}\cup\neg\{y\})\times P(\neg\{x\}\cup\{y\})}}{\sqrt{P(\{x\}\cup\{y\})\times P(\neg(\{x\}\cap\{y\}))}-\sqrt{P(\{x\}\cup\neg\{y\})\times P(\neg\{x\}\cup\{y\})}}$
κ	Cohen's	$\dfrac{P(\{x\}\cup\{y\})+P(\neg\{x\}\cup\neg\{y\})-P(\{x\})\times P(\{y\})-P(\neg\{x\})\times P(\neg\{y\})}{1-P(\{x\})\times P(\{y\})-P(\neg\{x\})\times P(\neg\{y\})}$
F	Certainty factor	$max\left(\dfrac{P(\{y\}|\{x\})-P(\{y\})}{1-P(\{y\})}, \dfrac{P(\{x\}|\{y\})-P(\{x\})}{1-P(\{x\})}\right)$

Table 10.7 Ranking of contingency tables using above interestingness measures

Example	ϕ	Q	Y	κ	F
E1	1	3	3	1	4
E2	2	1	1	2	1
E3	3	4	4	3	6
E4	4	2	2	5	2
E5	5	8	8	4	9
E6	6	7	7	7	7
E7	7	9	9	6	8
E8	8	10	10	8	10
E9	9	5	5	9	3
E10	10	6	6	10	5

Table 10.8 Ranking of contingency tables using overall influence

Example	Overall influence	Rank
E1	0.754	1
E2	0.627	3
E3	0.691	2
E4	0.560	4
E5	0.450	5
E6	0.352	7
E7	0.417	6
E8	0.167	9
E9	0.190	8
E10	0.023	10

10.4 Problem Statement

Let D be a database of customer transactions grown over a period of time. In this chapter, we are interested in making an influence analysis of a set of specific items. We will see how each of the specific items becomes influenced by different items in the database. One could view the entire database as a sequence of time-based (temporal) databases. For instance, such databases may concern consecutive years. To provide an incremental solution to this problem, one might need to mine only the current time database and combine the mining result with the previous mining results. Thus, one needs to mine only the current database for the purpose of making an analysis based on entire database. As a result one can obtain cost-effective and faster analysis based on the entire database. Since the database grows over time, an incremental solution to influence analysis of specific items becomes natural and desirable.

Each time database corresponds to the set of transactions made for a specific period of time. In this regard, the choice of time period corresponding to a database is an important issue. One could observe that the sales of items might vary over different seasons in a year. Instead of processing all the data together, we process data on a yearly basis. Then, the result of processing for the current year could be combined with that of previous years. Such incremental analysis might be appropriate since a season re-appears on a yearly basis. Otherwise, the processed result might be biased due to seasonal variations.

Our goal is to make an influence analysis of a set of items in a database. Let D_t be the database for the t-th period of time, $t = 1, 2, ..., n$. For computing overall influence between two items in a database, one needs to mine supports of itemsets of sizes one and two. The *size* of an itemset refers to the number of items in the itemset. Let $D_{1,k}$ be the collection of databases $D_1, D_2, ..., D_k$. For computing $OI(x, y, D_{1,k+1})$, we assume that $OI(x, y, D_{1,k})$ is available to us for items x, y in $D_{1,k}$. In other words, for computing $OI(x, y, D_{1,k+1})$, we have $supp(x, D_{1,k}), supp(y, D_{1,k})$, and $supp(x \cap y, D_{1,k})$. Thus, our incremental procedure needs to compute $supp(x, D_{1,k+1})$, $supp(y, D_{1,k+1})$, and $supp(x \cap y, D_{1,k+1})$ using (1) $supp(x, D_{1,k})$, $supp(y, D_{1,k})$, and $supp(x \cap y, D_{1,k})$, (2) $supp(x, D_{k+1}), supp(y, D_{k+1})$, and $supp(x \cap y, D_{k+1})$. In general, for an itemset X in $D_{1,k}$, $supp(X, D_{1,k+1})$ could be obtained incrementally as follows.

$$supp\left(X, D_{1,k+1}\right) = \frac{size(D_{k+1}) \times supp\left(X, D_{k+1}\right) + size\left(D_{1,k}\right) \times supp\left(X, D_{1,k}\right)}{size(D_{k+1}) + size\left(D_{1,k}\right)}$$

(10.4)

The $size(D)$ refers to the number of transactions in database D.

10.5 Related Work

For analyzing positive association between itemsets in a database, support-confidence framework was established by Agrawal et al. (1993). In Sect. 10.3.4, we have discussed why a confidence measure alone is not sufficient in determining an overall influence of an itemset on another itemset. Also, interestingness measures such as support, collective strength (Aggarwal and Yu 1998) and Jaccard (Tan et al. 2003) are not relevant in this context, since they are single-argument measures.

The χ^2 test (Greenwood and Nikulin 1996) only tells us whether two or more items are dependent. Such a test provides answers either "yes" or "no" to the question of whether the association is meaningful, and hence it might not be suitable for the specific requirement of our problem.

The interestingness measures such as lift (Tan et al. 2003), correlation (Tan et al. 2003), conviction (Brin et al. 1997), and odds-ratio (Tan et al. 2003) are semantically different from the measure of overall influence. Moreover, the values of each of these measures lie in $[0, \infty)$.

Shapiro (1991) has proposed leverage measure developed in the context of mining strong rules in a database. However, it might not be suitable for the specific requirement of our problem.

10.6 Design of Algorithms

Based on the discussion held in previous section, we design two algorithms for measuring influence of an item on another item and influence of an item on a set of specific items.

10.6.1 Designing Algorithm for Measuring Overall Influence of an Item on Another Item

In this algorithm (Adhikari and Rao 2010), we measure influence of an item on each of the items incrementally. We have expressed influence of an itemset on another itemset using supports of the relevant itemsets. Each itemset could be described by its *itemset* and *support*. We maintain arrays *IS1* and *IS2* for storing itemsets in $D_{1,k}$ of size one and two, respectively. *Itemset* attribute of ith itemset in *IS1* could be accessed using the notation $IS1(i).itemset$. Similar notation is used to access *support* attribute of an itemset. Also, we maintain arrays $\Delta IS1$ and $\Delta IS2$ for storing itemsets in D_{k+1} of size one and two, respectively. We merge *IS1* and $\Delta IS1$ to obtain supports of itemsets of size one in $D_{1,k+1}$ and are stored in array *OIS1*. Similarly, we merge *IS2* and $\Delta IS2$ to obtain supports of itemsets of size two in $D_{1,k+1}$ and are stored in array *OIS2*. Using *OIS1* and *OIS2*, we compute an overall influence between items in $D_{1,k+1}$. The overall influence between items is computed using formula (10.3) and stored in array *IOI*. The overall influence (*oi*) corresponding to j-th pair of items is accessed by $IOI(j).oi$.

Algorithm 10.1 Find top q overall influences in the given database over time.
procedure *Top-q-OI*(q, *IS1*, *IS2*, $\Delta IS1$, $\Delta IS2$, *IOI*)

Inputs:
q: an integer representing the number of top influences
IS1: array of supports of itemsets of size one in $D_{1,k}$
IS2: array of supports of itemsets of size two in $D_{1,k}$
$\Delta IS1$: array of supports of itemsets of size one in D_{k+1}
$\Delta IS2$: array of supports of itemsets of size two in D_{k+1}
Outputs:
IOI: array of overall influences in $D_{1,k+1}$
01: sort array $\Delta IS1$ on *itemset* attribute in non-decreasing order;
02: sort array $\Delta IS2$ on *itemset* attribute in non-decreasing order;
03: call *Merge* (*IS1*, $\Delta IS1$, *OIS1*);
04: call *Merge* (*IS2*, $\Delta IS2$, *OIS2*);
05: **let** $j = 1$;
06: **for** $i = 1$ to $|OIS2|$ **do**
07:      search $OIS2(i).item1$ in *OIS1*;
08:      search $OIS2(i).item2$ in *OIS1*;
09:      $IOI(j).oi = OI(OIS2(i).item1, OIS2(i).item2, D)$;
10:      $IOI(j).item1 = OIS2(i).item1; IOI(j).item2 = OIS2(i).item2$;
11:      increase j by 1;
12:      $IOI(j).oi = OI(OIS2(i).item2, OIS2(i).item1, D)$;
13:      $IOI(j).item1 = OIS2(i).item2; IOI(j).item2 = OIS2(i).item1$;
14:      increase j by 1;
15: **end for**
16: sort array *IOI* in non-increasing order on *oi* attribute;
17: return first q influences;
18: **end procedure**

The procedure *Merge* (A, B, C) merges sorted arrays A and B and generates output array C. In this context, sorting is based on support of an itemset. The time complexity of procedure *Merge* is $O(|A| + |B|)$ (Knuth 1998). Now, $OIS1$ contains the supports of items in $D_{1,k+1}$. Also, $OIS2$ contains the supports of itemsets of size two in $D_{1,k+1}$. The information contained in $OIS1$ and $OIS2$ is used to compute overall influence of an item on another item in $D_{1,k+1}$. Using line 09 we have computed influence of a singleton itemset on another singleton itemset. Suppose $\{6, 8\}$ be a frequent 2-itemset in $D_{1,k+1}$ stored in 4th cell of $OIS2$. Then $OI(OIS2(4).item1, OIS2(4).item2, D_{1,k+1})$ refers to overall influence of $\{6\}$ on $\{8\}$ in $D_{1,k+1}$. In lines 6–15, we have computed and stored overall influences of a singleton itemset on another singleton itemset in $D_{1,k+1}$. In line 16, we have sorted overall influences in non-increasing order. Finally, we display first q overall influences.

Let $IS1$ and $IS2$ contain M and N itemsets respectively. Let $\Delta IS1$ and $\Delta IS2$ contain m and n elements respectively. Lines 1 and 2 take $O(m \times log(m))$ and $O(n \times log(n))$ time, respectively. Also, lines 3 and 4 take $O(M + m)$ and $O(N + n)$ time, respectively. Each of the search statements in lines 7 and 8 take $O(log(M + m))$ time, since $OIS1$ is sorted. The sort statement in line 16 takes $O((N + n) \times log(N + n))$. The time complexity of lines 6–15 is $O((N + n) \times log(M + m))$. Thus, the time complexity of algorithm *Top-q-OI* is *maximum* $\{O(M + m), O((N + n) \times log(N + n)), O((N + n) \times log(M + m))\}$.

10.6.2 Designing Algorithm for Measuring Overall Influence of an Item on Each of the Specific Items

One could store specific items in an array. The proposed algorithm seems to be the same as Algorithm 10.1 except that every time it measures an overall influence of an item on a specific item.

10.6.3 Designing Algorithm for Identifying Top Influential Items on a Set of Specific Items

In Algorithm 10.2 (Adhikari and Rao 2010), we find influence of an item on a set of specific items in a database. We construct influence matrix (IM) from the arrays of specific items (SI) and overall influence between items (IOI). The algorithm scans IM for each item to count the number of interesting influences, which are stored in array called *count*. Finally, we sort *count* in descending order based on primary key count value and secondary key support.

Algorithm 10.2 Find influence of an item on a set of specific items in the database over time.
procedure *Top-q-items*(q, SI, $IS1$, $IS2$, $\Delta IS1$, $\Delta IS2$, $OIS1$, $OIS2$, IOI)

Inputs:
q: an integer representing the number of top influences
SI: array of specific items
IS1, IS2, ΔIS1, ΔIS2, OIS1, OIS2, IOI: as specified in Algorithm 10.1
Outputs:
count: array of number of interesting influences
01: **for** $i = 1$ to $|SI|$ **do**
02:   **for** $j = 1$ to $|IOI|$ **do**
03:     **if** $(SI(i) = IOI(j).item1)$ **then**
04:       $IM(i)(j) = IOI(j).oi$;
05:     **end if**
06:   **end for**
07: **end for**
08: **for** $j = 1$ to $|IOI|$ **do**
09:   **let** $count(j) = 0$;
10:   **for** $i = 1$ to $|SI|$ **do**
11:     **if** $(IM(j)(i) \geq \gamma)$ **then**
12:       increase $count(j)$ by 1;
13:     **end if**
14:   **end for**
15: **end for**
16: sort *count* on non-increasing order on primary key count value and secondary key support;
17: return first *q* items;
end procedure

Let array *SI* contains *p* items. Line 1 repeats for *p* times. Line 2 repeats $O(M + m)$ times. So, lines 1–7 take $O(p \times (M + m))$ times. Line 8 repeats $O(M + m)$ times. Line 10 repeats *p* times. Thus, line 8–15 take $O(p \times (M + m))$ times. Therefore, the time complexity of the above algorithm is $O(p \times (M + m))$, where $M > m$. Also, sorting statement at line 16 takes $O((M + m) \times log(M + m))$. The time complexity of algorithm *Top-q-items* is $maximum\{O(p \times (M + m)), O((M + m) \times log(M + m))\}$.

10.7 Experiments

We have carried out several experiments to study the effectiveness of the proposed analysis. All the experiments have been implemented on a 1.6 GHz Pentium IV with 256 MB of memory using Visual C++ (version 6.0) software. We present the experimental results using three real-world databases and one synthetic database. The databases *mushroom*, *retail* (Frequent itemset mining dataset repository) and *ecoli* are real-world databases. Database *ecoli* is a subset of *ecoli database* (UCI ML repository) and it has been processed for the purpose of conducting experiments. *Random-68* is a synthetic database. The symbols used in different tables are explained as follows. Let *D*, *NT*, *ALT*, *AFI*, and *NI* denote database, the number of transactions, average length of a transaction, average frequency of an item, and number of items, respectively. The details of these databases are given in Table 10.9.

Table 10.9 Database characteristics

Database	NT	ALT	AFI	NI
Mushroom (M)	8,124	24.000	1,624.800	120
E. coli (E)	336	7.000	25.835	91
Random-68 (R)	3,000	5.460	280.985	68
Retail (Rt)	88,162	11.306	99.674	10,000

Each database has been divided into 10 databases, called input databases, for the purpose of conducting experiments on multiple time databases. The input databases obtained from *mushroom, E.coli, random-68* and *retail* are named as M_i, E_i, R_i, and Rt_i, $i = 0, 1, ..., 9$. We present some characteristics of the input databases, see (Tables 10.10). Top 10 overall influences are presented in Table 10.11.

We have studied the execution time with respect to the number of data sources. We observe in Figs. 10.1, 10.2, 10.3 and 10.4 that this time increases as the number of data sources gets higher.

The size of each input database generated from *mushroom* and *retail* are significantly larger than an input database generated from *E.coli*. As a result, we observe a steeper relationship in Figs. 10.1 and 10.4. The number of frequent itemsets decreases as the minimum support increases.

In Figs. 10.5, 10.6, 10.7 and 10.8 it is shown how the execution time decreases over the increase of the minimum support value.

Table 10.10 Time database characteristics

D	NT	ALT	AFI	NI	D	NT	ALT	AFI	NI
M_0	812	24.000	295.273	66	M_5	812	24.000	221.454	88
M_1	812	24.000	286.588	68	M_6	812	24.000	216.533	90
M_2	812	24.000	249.846	78	M_7	812	24.000	191.059	102
M_3	812	24.000	282.435	69	M_8	812	24.000	229.271	85
M_4	812	24.000	259.840	75	M_9	816	24.000	227.721	86
E_0	33	7.000	4.620	50	E_5	33	7.000	3.915	59
E_1	33	7.000	5.133	45	E_6	33	7.000	3.500	66
E_2	33	7.000	5.500	42	E_7	33	7.000	3.915	59
E_3	33	7.000	4.813	48	E_8	33	7.000	3.397	68
E_4	33	7.000	3.397	68	E_9	39	7.000	4.550	60
R_0	300	5.590	28.676	68	R_5	300	5.140	26.676	68
R_1	300	5.417	28.000	68	R_6	300	5.510	28.353	68
R_2	300	5.360	27.647	68	R_7	300	5.497	28.338	68
R_3	300	5.543	28.456	68	R_8	300	5.537	28.471	68
R_4	300	5.533	28.382	68	R_9	300	5.477	28.235	68
Rt_0	9,000	11.244	12.070	8,384	Rt_5	9,000	10.856	16.710	5,847
Rt_1	9,000	11.209	12.265	8,225	Rt_6	9,000	11.200	17.416	5,788
Rt_2	9,000	11.337	14.597	6,990	Rt_7	9,000	11.155	17.346	5,788
Rt_3	9,000	11.490	16.663	6,206	Rt_8	9,000	11.997	18.690	5,777
Rt_4	9,000	10.957	16.039	6,148	Rt_9	7,162	11.692	15.348	5,456

Table 10.11 Top 10 overall influences in different databases

M (supp = 0.15)			E (supp = 0.12)			R (supp = 0.03)			Rt (supp = 0.12)		
{x}	{y}	OI	{x}	{y}	OI	{x}	{y}	OI	{x}	{y}	OI
86	34	0.997	24	48	0.946	19	29	−0.017	41	39	0.200
34	86	0.992	89	50	0.913	29	19	−0.020	39	48	0.180
58	24	0.991	53	48	0.693	8	56	−0.023	48	39	0.175
67	76	0.986	63	50	0.665	56	8	−0.023	41	48	0.129
76	67	0.986	87	50	0.660	15	14	−0.031	39	41	0.114
24	58	0.963	56	50	0.621	14	15	−0.032	48	41	0.071
93	59	0.895	61	50	0.618	18	52	−0.035	48	7	−0.234
93	76	0.884	27	48	0.618	52	18	−0.036	39	7	−0.292
93	67	0.881	83	50	0.540	54	58	−0.044	48	2	−0.293
102	24	0.875	56	48	0.488	58	54	−0.047	48	1	−0.316

Fig. 10.1 Execution time versus number of databases at *supp* = 0.2 (*mushroom*)

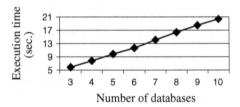

Fig. 10.2 Execution time versus number of databases at *supp* = 0.12 (*E. coli*)

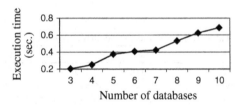

Fig. 10.3 Execution time versus number of databases at *supp* = 0.03 (*random-68*)

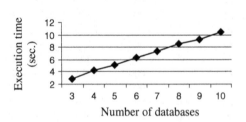

Fig. 10.4 Execution time versus number of databases at *supp* = 0.2 (*retail*)

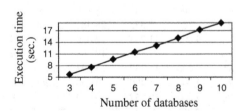

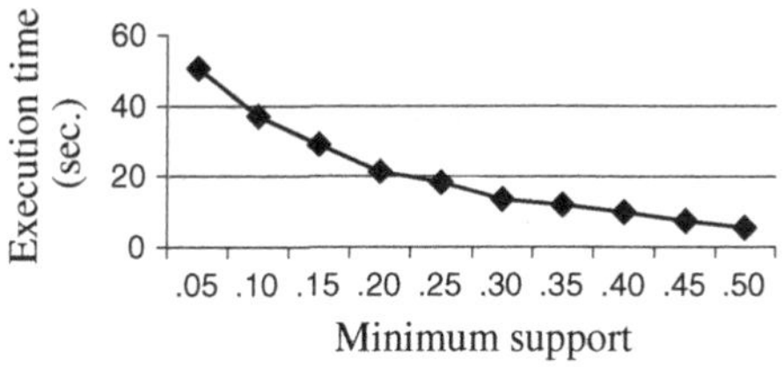

Fig. 10.5 Execution time
versus minimum support
(*mushroom*)

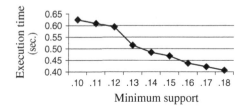

Fig. 10.6 Execution time
versus minimum support
(*E. coli*)

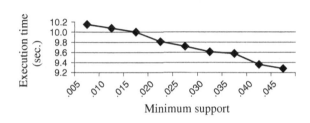

Fig. 10.7 Execution time
versus minimum support
(*random-68*)

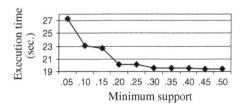

Fig. 10.8 Execution time
versus minimum support
(*retail*)

By comparing Figs. 10.1, 10.2, 10.3 and 10.4, one notes that the steepness of a graph increases as the size of branch databases increases. Similar observation holds true on Figs. 10.5, 10.6, 10.7 and 10.8.

In Sect. 10.3.1 we have explained the concept of interesting overall influence. Given a threshold value of γ, we have counted the number of overall influences. In Figs. 10.9, 10.10, 10.11, and 10.12 we have shown how the number of interesting overall influence decreases over the increase of the minimum influence level.

Figures 10.9, 10.10, 10.11 and 10.12 also provide another type of insight. As the size of a transaction increases, the number of interesting overall influences also increases, provided the number of transactions in a branch database and the level of overall influence remain constant. The average transaction length of mushroom branch databases is significantly higher than that of other branch databases.

Fig. 10.9 Number of interesting *OI* values versus γ at *supp* = 0.2 (*mushroom*)

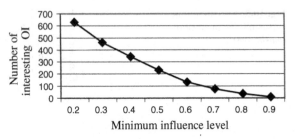

Fig. 10.10 Number of interesting *OI* values versus γ at *supp* = 0.12 (*E. coli*)

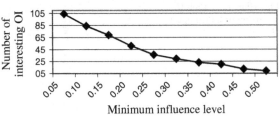

Fig. 10.11 Number of interesting *OI* values versus γ at *supp* = 0.015 (*random-68*)

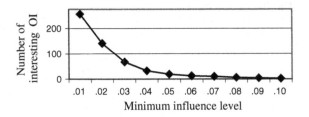

Fig. 10.12 Number of interesting *OI* values versus γ at *supp* = 0.02 (*retail*)

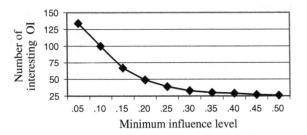

The mining algorithm generates a large number of interesting overall influences even at the minimum influence level of 0.2.

We have taken specific items in different databases in Table 10.12. Based on the requirement of association analysis one could choose specific items in time databases.

Table 10.12 Specific items in different databases

M	E	R	Rt
SI = {1, 2, 3, 6, 9, 10, 11, 13, 16, 23}	SI = {37, 39, 40, 41, 42, 44, 48, 49, 50, 51}	SI = {1, 2, 3, 4, 5, 6, 7, 8, 9, 10}	SI = {0, 1, 2, 3, 4, 5, 6, 7, 8, 9}

Table 10.13 Influences of different items on a set of specific items in different databases

M (supp = 0.2)		E (supp = 0.12)		R (supp = 0.015)		Rt (supp = 0.03)	
γ	$x(\eta)$	γ	$x(\eta)$	γ	$x(\eta)$	γ	$x(\eta)$
0.3	86(3), 34(3), 36(3), 39(2), 59(2), 63(2), 2(2), 93(2), 36(2), 23(2), 90(1), 24(1)	0.07	48(5), 37(2), 50(1), 42(1), 44(1), 39(1), 40(1), 49(1), 41(1)	0.05	18(5), 15(3), 65(2), 55(2), 61(2), 7(1), 54(1), 27(1), 35(1), 66(1), 22(1)	0.05	413(8), 310(2), 0(1), 1(1), 8(1), 2(1), 3(1),5(1), 9(1), 4(1)

The influences of different items on a set of specific items in different databases are presented in Table 10.13. In the *mushroom* database, item 86 is the most influential item because 3 specific items are influenced by it. Item 48 in *E. coli* database exhibits a significant influence on the set of specific items. It shows that item 48 has high influence on 5 out of 10 specific items. In the same way one could conclude that item 18 in *random-68* is the most influential item with respect to the given set of specific items. Five out of 10 specific items are influenced significantly by item 18. Item 413 influences 8 out of 10 specific items significantly in *retail* database. Therefore, it is the most influential item in *retail*.

10.8 Conclusions

The concept of positive influence might not be sufficient in many data analyses. One could perform an effective data analysis by using the measure of overall influence. Measuring influence over time becomes an important issue, since many companies possess data over a long period of time so that they could be exploited in an efficient manner. In this chapter, we have designed two algorithms using the measure of overall influence. The first algorithm reports all the significant influences in a database. In the second algorithm, we have sorted items based on their influences on a set of specific items. Such analyses might be interesting since the proposed measure of influence considers both positive and negative influence of an itemset on another itemset.

References

Adhikari A, Rao PR (2007) A framework for mining arbitrary Boolean expressions induced by frequent itemsets. In: Proceedings of the international conference on artificial intelligence, pp 5–23

Adhikari A, Rao PR (2008a) Efficient clustering of databases induced by local patterns. Decis Support Syst 44(4):925–943

Adhikari A, Rao PR (2008b) Mining conditional patterns in a database. Pattern Recogn Lett 29(10):1515–1523

Adhikari A, Rao PR (2008c) Synthesizing heavy association rules in different real data sources. Pattern Recogn Lett 29(1):59–71

Adhikari J, Rao PR (2010) Measuring influence of an item in a database over time. Pattern Recogn Lett 31(3):179–187

Adhikari J, Rao PR, Adhikari A (2009) Clustering items in different data sources induced by stability. Int Arab J Inf Technol 6(4):66–74

Adhikari A, Ramachandrarao P, Pedrycz W (2011) Study of select items in different data sources by grouping. Knowl Inf Syst 27(1):23–43

Aggarwal C, Yu P (1998) A new framework for itemset generation. In: Proceedings of PODS, pp 18–24

Agrawal R, Srikant R (1994) Fast algorithms for mining association rules. In: Proceedings of the international conference on very large databases, pp 487–499

Agrawal R, Imielinski T, Swami A (1993) Mining association rules between sets of items in large databases. In: Proceedings of the ACM SIGMOD conference management of data, pp 207–216

Brin S, Motwani R, Ullman JD, Tsur S (1997) Dynamic itemset counting and implication rules for market basket data. In: Proceedings of the ACM SIGMOD international conference on management of data, pp 255–264

Gary JR, Petersen A (2000) Analysis of cross category dependence in market basket selection. J Retail 76(3):367–392

Greenwood PE, Nikulin MS (1996) A guide to chi squared testing. Wiley, Hoboken

Han J, Pei J, Yiwen Y (2000) Mining frequent patterns without candidate generation. In: Proceedings of the ACM-SIGMOD international conference management of data, pp 1–12

Knuth DE (1998) The art of computer programming, sorting and searching, vol 3. Addison-Wesley, Boston

Savasere A, Omiecinski E, Navathe S (1995) An efficient algorithm for mining association rules in large databases. In: Proceedings of the international conference on very large data bases, pp 432–443

Shapiro P (1991) Discovery, analysis, and presentation of strong rules. In: Proceedings of knowledge discovery in databases, pp 229–248

Tan PN, Kumar V, Srivastava J (2003) Selecting the right interestingness measure for association patterns. In: Proceedings of SIGKDD conference, pp 32–41

Wu X, Zhang C, Zhang S (2004) Efficient mining of both positive and negative association rules. ACM Trans Inf Syst 22(3):381–405

Wu X, Zhang C, Zhang S (2005) Database classification for multi-database mining. Inf Syst 30(1):71–88

Frequent itemset mining dataset repository, http://fimi.cs.helsinki.fi/data

UCI ML repository, http://www.ics.uci.edu/~mlearn/MLSummary.html

Chapter 11
Summary and Conclusions

Recognition of patterns and associations in a large database is a natural, interesting, timely and practically relevant activity. It becomes more interesting as well as challenging when we are required to identify patterns and associations in multiple large data sources. While dealing with the domain of multiple large data sources, it has been observed that many patterns are specific to this domain; also some patterns are extensions of classical patterns. In this book, we have discussed recent work in mining patterns and associations in multiple large data sources, and presented some chapter-wise conclusions. While mining multiple large databases challenges are increased manifold. Yet, the challenges of mining a large database have not ended. Thus, mining and interpreting patterns and associations become a more complex issue. In this concluding section, we list a number of such essential challenges.

11.1 Changing Scenarios

When there are multiple databases originated from different sources, we may require applying various data analyses including extraction of useful patterns from these data sources. In this book, we have presented recent advancements in extracting patterns and associations in the domain of multiple large databases. Data mining methods and data analyses dealing with classical patterns, such as association rules (Agrawal et al. 1993), intervals (Allen 1983), sequential patterns (Agrawal and Srikant 1995), and episodes (Mannila et al. 1995) dominated at the beginning of KDD research. Over the time, organizations are required to deal with different datasets across various domains. As a result, new techniques of data mining have emerged (Adhikari et al. 2010a; Jin et al. 2008) to deal with multiple large databases; classical patterns have appeared in specialized forms (Wu and Zhang 2003; Adhikari and Rao 2008a); new patterns and associations (Adhikari et al. 2011b; Adhikari and Rao 2008b) have also been reported subsequently. Mining multiple data sources is a strategically important area, since one could obtain multiple databases from different domains. Many applications of multi-database mining have

A. Adhikari et al., *Data Analysis and Pattern Recognition in Multiple Databases*,
Intelligent Systems Reference Library 61, DOI: 10.1007/978-3-319-03410-2_11,
© Springer International Publishing Switzerland 2014

been reported from different domains such as market basket data (Adhikari et al. 2010b), biological data (Knobbe 2004; Chen and Lonardi 2009) and privacy preserving data (Aggarwal and Yu 2008).

11.2 Summary of Chapters

It may be the case that the collection of multiple databases could be very large. Thus, most of the multi-database mining techniques become approximate in nature. In Chap. 1, we have discussed three approaches to mining multiple large databases. While dealing with multiple databases, one may have to apply various data preprocessing techniques such as preparation of different data warehouses, selection of appropriate databases and ordering of databases. Pattern recognition and association analysis are the two important tasks of knowledge discovery process. Some patterns such as high frequency association rule, heavy association rule and exceptional pattern, are specific to multi-database mining. Multi-database mining is an important area of research since many applications concern data distributed over multiple sources.

Association rule mining is an important task of data mining. The collection of databases from various sources is often very large. In many cases either it is difficult to move data at one place or data movement is not necessary. Sometimes there could be some privacy issues that have to be taken into consideration. Local pattern analysis (Zhang et al. 2003) becomes then a useful solution to multi-database mining. In Chap. 2, we have presented an extended model of local pattern analysis. In the context of multi-database mining, we often come across with various extreme patterns, and synthesizing global patterns remains a critical issue. We have presented an algorithm to synthesize high frequency association rule, heavy association rule and exceptional association rule in multiple databases (Adhikari and Rao 2008a).

Chapter 3 deals with clustering items in multiple databases based on stability of an item. In this chapter, we have defined stability of an item based on its supports over time (Adhikari et al. 2009). Also, we have defined the best class of items, where the variation among the items is the least. Stable items are useful in making strategic decisions for an organization.

Global patterns are useful for global data analyses as well as global decision making problems. There have been some difficulties in merging all the local databases and finding global patterns in multiple databases. In Chap. 4, we present pipelined feedback technique for mining global patterns in multiple databases (Adhikari et al. 2010a). Although this technique mines approximate global patterns in multiple databases, but it improves significantly the quality of global patterns over the previous techniques. When many databases are mined for local patterns individually, the mined patterns from the previous databases are also given as input to mining of current database. Thus, such feedback mechanism improves the quality of patterns in multiple databases.

A multi-branch company might deal with several items through different branches. All the local transactions are stored locally. We have proposed a measure A_2 for capturing association among items in a database (Adhikari and Rao 2008b). Based on the association among items, we have presented an algorithm for clustering items in multiple databases (Adhikari 2013). The algorithm is based on synthesized support of an itemset in multiple databases. In Chap. 5, we have also presented algorithm for synthesizing high frequency itemsets in multiple databases (Adhikari 2013).

In Chap. 6, we are interested in patterns of select items in multiple databases. It might be judicious to mine subset of entire dataset in view of searching those patterns. For this purpose, we have presented a model of mining global patterns of select items from multiple databases (Adhikari et al. 2011a). We have presented a measure for calculating overall association between two items in a database. This measure has been used to group the items in multiple databases. Each group grows around a select item. An item falls in a group, if the overall association between the item and the nucleus item in the group is significantly more. We have presented an algorithm to identify such groups (Adhikari et al. 2011a). It helps to study the influence of select items in multiple databases.

Finding global exceptional patterns in a database is an interesting issue. In Chap. 7, we have presented different types of exceptional patterns in multiple databases. In the context of multiple databases, there are two types of exceptional patterns viz., type I and II. A type II exceptional pattern is heavily supported by a few databases, but remains absent in most of the databases. On the other hand, a type I exceptional pattern is supported by most of the databases. Yet, it does not possess heavy support in multiple databases. We have presented an algorithm to identify type II exceptional patterns in multiple databases (Adhikari 2012).

Many large organizations collect data over a long period of time. Such data might exhibit some exceptionality. Before identifying exceptional patterns in a database constructed over a long period of time, we identify a basic change in sales data, called notch. It is an increase in sales in a year followed by decrease in sales in the next year or vice versa. We have generalized the concept of notch, giving rise to a generalized notch. Some generalized notches that satisfy user-defined height and width criteria are called icebergs. In Chap. 8, we have presented an algorithm to identify icebergs in time stamped databases (Adhikari et al. 2011b).

In Chap. 9, we have presented an algorithm for identifying calendar-based periodic patterns in time-stamped databases. In a previous work, authors introduced the concept of certainty factor in association with an overlapped interval. We have presented an improved method by incorporating support information for effective analysis of overlapped intervals. We have introduced a hash based data structure for storing and managing patterns. We have presented an algorithm that identifies both full and partial periodic calendar-based patterns in time-stamped databases (Adhikari and Rao 2013).

The concepts influence and association are different. In Chap. 10, we have studied the influence of an itemset on another itemset and presented a measure of influence (Adhikari and Rao 2010). Also, we have studied the properties of influence measure. We have presented two algorithms using the measure of overall influence (Adhikari and Rao 2010). The first algorithm reports all the significant influences in a database, where as the second one reports the sorted influences on a set of specific items.

11.3 Selected Open Problems and Challenges

We discuss here a few open problems. Multi-database mining has remained an unexplored area of research. A few problems are discussed here; those could motivate further research.

1. Designing and querying data warehouse for multi-database mining is an important issue. Development of a multi-databases mining system would depend on these aspects. It could be an important topic in designing a decision support system that deals with discovering knowledge from multiple databases.
2. The roles of sampling theory remains important while mining multiple large databases. Many a times, it is not possible to mine multiple very large databases. Sampling theory could play an important role in approximately mining patterns in multiple large databases. Various strategies of sampling could play an important role in mining good quality of patterns in multiple large databases.
3. Mining multiple stream data is a challenging problem. One important characteristic is that a stream data never ends. Sometimes stream data are collected from multiple sensors. Mining multiple data streams could be another problem of research (Wu and Gruenwald 2010).

11.4 Conclusions

Most of the patterns searched in a single database, are also applicable to multiple databases. Thus, classical patterns such as classification, clustering, and association rules are also patterns of interest in multiple databases. Some interesting patterns such as heavy association rule, global exceptional patterns and high frequency rule are specific to a multi-database mining system.

Currently, most of the data mining algorithms generate patterns that can be defined in a mathematical sense. However, methods for explaining the meaning of the found patterns are still not sufficient. In the light of the increased object complexity, this problem will gain additional importance. Thus, it is likely that not only the input data for data mining is getting more complex, but also the output patterns will increase in complexity. This trend is amplified by another challenge that recently came up in the data mining community. In many applications, the very general patterns derived by the standard methods do not yield a satisfying solution to the given task. In order to solve a problem, the found patterns need to fulfill a certain set of constraints which make them more interesting for the application. Examples of this type of patterns are correlation clusters (Bohm et al. 2004) and constrained association rules (Srikant et al. 1997). For more complex data with mutual relationships, the derived patterns will be even more complex. Thus, one needs to address the following additional challenges (Kriegel et al. 2007).

- The patterns described by the data mining algorithms are still too abstract for being understood. However, a pattern that is misinterpreted is of great danger. For example, many data mining algorithms do not distinguish between causality and co-occurrence. Consider an application that aims at finding the reason for a certain type of disease. There is a great difference between finding the origin of the disease or finding just an additional symptom. Therefore, the following old challenge will remain an important issue for the data mining community: Developing systems which derive understandable patterns and making already derived patterns understandable.
- Current algorithms of data mining mostly focus on a limited set of standard patterns. However, deriving these patterns often does not yield a direct and complete solution to many problems, where data mining could be very useful. Furthermore, with an increasing complexity of the analyzed data, it is likely that the derived patterns will increase in the complexity as well. Thus, a future trend in data mining will be of finding richer patterns.
- In view of finding future patterns, we need to increase the number of valid patterns, and it could be done through trying out several parameter values while dealing with a large data set of complex objects. Therefore, the number of potentially valid patterns could be too large to handle if an automatic system is not employed. Thus, future systems must provide a platform for pattern exploration, where users can browse the knowledge that they consider interesting. In conclusion, future data mining should generate a large variety of well understandable patterns. Due to variations to the parameterization schemes, the number of possibly meaningful and useful patterns would dramatically increase. Therefore, it is important to manage and visualize these patterns efficiently.

Patterns and data analyses presented in this book are just a beginning for multiple related databases. We believe that many more patterns as well as data analyses are yet to come in a multi-database mining environment.

References

Adhikari A (2012) Synthesizing global exceptional patterns in different data sources. J Intell Syst 21(3):293–323

Adhikari A (2013) Clustering local frequency items in multiple databases. Inf Sci 237:221–241

Adhikari A, Rao PR (2008a) Synthesizing heavy association rules from different real data sources. Pattern Recogn Lett 29(1):59–71

Adhikari A, Rao PR (2008b) Capturing association among items in a database. Data Knowl Eng 67(3):430–443

Adhikari J, Rao PR (2010) Measuring influence of an item in a database over time. Pattern Recogn Lett 31(3):179–187

Adhikari J, Rao PR (2013) Identifying calendar-based periodic patterns. In: Ramanna S, Jain LC, Howlett RJ (eds) Emerging paradigms in machine learning. Springer, Berlin, pp 329–357

Adhikari J, Rao PR, Adhikari A (2009) Clustering items in different data sources induced by stability. Int Arab J Inf Technol 6(4):394–402

Adhikari A, Rao PR, Prasad B, Adhikari J (2010a) Mining multiple large data sources. Int Arab J Inf Technol 7(2):243–251

Adhikari A, Ramachandrarao P, Pedrycz W (2010b) Developing multi-databases mining applications. Springer, New York

Adhikari A, Ramachandrarao P, Pedrycz W (2011a) Study of select items in different data sources by grouping. Knowl Inf Syst 27(1):23–43

Adhikari J, Rao PR, Pedrycz W (2011b) Mining icebergs in time-stamped databases. In: Indian international conferences on artificial intelligence. pp 639–658

Aggarwal CC, Yu PS (2008) Privacy-preserving data mining—models and algorithms. Springer, New York

Agrawal R, Srikant R (1995) Mining sequential patterns. In: Proceedings of the 11th international conference on data engineering. pp 3–14

Agrawal R, Imielinski T, Swami A (1993) Mining association rules between sets of items in large databases. In: Proceedings of ACM SIGMOD conference. pp 207–216

Allen JF (1983) Maintaining knowledge about temporal intervals. Commun ACM 26(11):832–843

Bohm C, Kailing K, Kröger P, Zimek A (2004) Computing clusters of correlation connected objects. In: Proceedings of the SIGMOD conference 2004. pp. 455–466

Chen JY, Lonardi S (2009) Biological data mining. CRC Press, Boca Raton

Jin Y, Murali TM, Ramakrishnan N (2008) Compositional mining of multirelational biological datasets. ACM Trans Knowl Discov Data 2(1):1–32

Knobbe AJ (2004) Multi-relational data mining. Ph.D. thesis, Aan de Universiteit Utrecht, Netherlands

Kriegel HP, Borgwardt KM, Kröger P, Pryakhin A, Schubert M, Zimek A (2007) Future trends in data mining. Data Min Knowl Disc 15:87–97

Mannila H, Toivonen H, Verkamo I (1995) Discovery of frequent episodes in event sequences. In: Proceedings of the 1st international conference on knowledge discovery and data mining. pp 210–215

Srikant R, Vu Q, Agrawal R (1997) Mining association rules with item constraints. In: Proceedings of the 3rd ACM international conference on knowledge discovery and data mining (KDD). pp 67–73

Wu W, Gruenwald L (2010) Research issues in mining multiple data streams. In: StreamKDD'10, ACM international conference on knowledge discovery and data mining. pp 56–60

Wu X, Zhang S (2003) Synthesizing high-frequency rules from different data sources. IEEE Trans Knowl Data Eng 14(2):353–367

Zhang S, Wu X, Zhang C (2003) Multi-database mining. IEEE Comput Intell Bull 2(1):5–13

Index

A. Adhikari et al., *Data Analysis and Pattern Recognition in Multiple Databases*, Intelligent Systems Reference Library 61, DOI: 10.1007/978-3-319-03410-2, © Springer International Publishing Switzerland 2014

Printed in the United States
By Bookmasters